Lecture Notes in Economics and Mathematical Systems 632

For further volumes:
http://www.springer.com/series/300

Alexander Saichev • Yannick Malevergne
Didier Sornette

Theory of Zipf's Law and Beyond

Springer

Professor Alexander Saichev
Mathematical Department
Nizhni Novgorod State University
Gagarin Prospekt 23
603950 Nizhni Novgorod
Russia
saichev@hotmail.com

Professor Didier Sornette
Department of Management
Technology and Economics
ETH Zürich
Kreuzplatz 5
8032 Zurich
Switzerland
dsornette@ethz.ch

Professor Yannick Malevergne
ISEAG
University of Saint-Etienne
2 rue Tréfilerie
42023 Saint-Etienne cedex 2
France

and

EMLYON Business School – Cefra
23 avenue Guy de Collongue
69134 Ecully Cedex
France
ymalevergne@ethz.ch

ISSN 0075-8442
ISBN 978-3-642-02945-5 e-ISBN 978-3-642-02946-2
DOI 10.1007/978-3-642-02946-2
Springer Heidelberg Dordrecht London New York

Library of Congress Control Number: 2009932422

Cover design: SPi Publisher Services

Printed on acid-free paper

Springer is part of Springer Science+Business Media (www.springer.com)

Preface

Zipf's law is one of the few quantitative reproducible regularities found in economics. It states that, for most countries, the size distributions of cities and of firms (with additional examples found in many other scientific fields) are power laws with a specific exponent: the number of cities and firms with a size greater than S is inversely proportional to S. Most explanations start with Gibrat's law of proportional growth but need to incorporate additional constraints and ingredients introducing deviations from it. Here, we present a general theoretical derivation of Zipf's law, providing a synthesis and extension of previous approaches. First, we show that combining Gibrat's law at all firm levels with random processes of firm's births and deaths yield Zipf's law under a "balance" condition between a firm's growth and death rate. We find that Gibrat's law of proportionate growth does not need to be strictly satisfied. As long as the volatility of firms' sizes increase asymptotically proportionally to the size of the firm and that the instantaneous growth rate increases not faster than the volatility, the distribution of firm sizes follows Zipf's law. This suggests that the occurrence of very large firms in the distribution of firm sizes described by Zipf's law is more a consequence of random growth than systematic returns: in particular, for large firms, volatility must dominate over the instantaneous growth rate. We develop the theoretical framework to take into account (1) time-varying firm creation, (2) firm's exit resulting from both a lack of sufficient capital and sudden external shocks, (3) the coupling between firm's birth rate and the growth of the value of the population of firms. We predict deviations from Zipf's law under a variety of circumstances, for instance, when the balance between the birth rate, the instantaneous growth rate and the death rate is not fulfilled, providing a framework for identifying the possible origin(s) of the many reports of deviations from the pure Zipf's law. Reciprocally, deviations from Zipf's law in a given economy provides a diagnostic, suggesting possible policy corrections. The results obtained here are general and provide an underpinning for understanding and quantifying Zipf's law and the power law distribution of sizes found in many fields.

Nizhni Novgorod, Russia — A. Saichev
Lyon & Saint-Etienne, France — Y. Malevergne
Zürich, Switzerland — D. Sornette

Contents

Symbols

$a(s)$	Drift of the asset value process; In the case of a geometric Brownian motion, $a(s) = a \cdot s$, so that a denotes the instantaneous rate of return
$b(s)$	Volatility of the asset value process; In the case of a Geometric Brownian Motion, $b(s) = b \cdot s$
c	Drift of the log-asset value process for the Geometric Brownian Motion. The three parameters a, b and c are related by $a = c + \frac{b^2}{2}$ (see 2.14)
d	Exponent of exponentially growing intensity of births
$f(s;t)$	Probability density function of a single firm's size (see 3.13)
$f_d(t)$	Probability density function of the life duration above a given size level (see 5.8)
$f_k(t)$	Probability density function of the life duration when firms exit upon first reaching a given size level from above (see 5.19)
$F(s;t)$	Cumulative distribution function of a single firm's size (see 3.6)
$\bar{F}(s;t)$	Complementary cumulative distribution function of a single firm's size
$\bar{F}_d(t)$	Complementary cumulative distribution of the life duration above a given size level (see 5.16)
$\bar{F}_k(t)$	Complementary cumulative distribution function of the life duration when firms exit upon first reaching a given size level from above (see 5.5)
$g(s)$	Steady-state mean density of firm's size (see Definition 3.3.1)
$g(s,t)$	Mean density of the size of all incumbent firms at time t (see 3.12)
$G(s;t)$	Mean number of all incumbent firms of size larger than s
m	Tail index of the distribution of firm's size
$M(s-s',s')$	Rate of merger and acquisition (M&A) between firms of size $s-s'$ and s' to create a new firm of size s
$SO(s-s',s')$	Rate of creation of spin-off firms of size s' from a firm of size s which retains a value $s-s'$ after the spin-off creation
$\mathcal{Q}(t)$	Probability that the life-span of a firm is larger than t
S	Firm's size / asset value
s_0	Initial firm's size

s_1	Level of firm's killing (death)
t_b	Characteristic time associated with the volatility $t_b = 2/b^2$
$W(t)$	Standard Wiener process
$X(t, c, b)$	Geometric Brownian motion (2.11)
$Y(t, c, b)$	Wiener process with drift (2.6)
$\Omega(t)$	Overall mean asset value of the economy at time t
λ	Growth to risk ratio, $\lambda = \frac{2c}{b^2}$ (2.19)
δ	Return to risk ratio, $\delta = \frac{2a}{b^2} = 1 + \lambda$ (6.111)
$\delta(s)$	Generalization of δ to $\delta(s) := \frac{2a(s)}{b^2(s)}$ (6.94)
δ'	Inverse of δ: $\delta' := \frac{1}{1+\lambda} = \frac{b^2}{2a}$ (7.45)
ϕ	Density of the Gaussian distribution for the Wiener process with drift $Y(t, c, b)$ (see 2.8)
μ	Hazard rate of firm's sudden death
ν	Intensity of firm's birth (see Assumption 3)
ψ	Density of the log-normal distribution (2.32)
τ	Reduced time, $\tau = \frac{b^2 t}{2}$
ζ	Hazard rate to risk ratio $\zeta := \frac{2\mu}{b^2}$ (7.13)
$\mathbf{1}(x)$	Indicator function of the event $x \geqslant 0$

Chapter 1
Introduction

One of the broadly accepted universal laws of complex systems, particularly relevant in social sciences and economics, is that proposed by Zipf (1949). Zipf's law usually refers to the fact that the probability $P(s) = \Pr\{S > s\}$ that the value S of some stochastic variable, usually a size or frequency, is greater than s, decays with the growth of s as $P(s) \sim s^{-1}$. This in turn means that the probability density functions $p(s)$ exhibits the power law dependence

$$p(s) \sim 1/s^{1+m} \quad \text{with } m = 1. \tag{1.1}$$

Perhaps the distribution most studied from the perspective of Zipf's law is that of firm sizes, where size is proxied by sales, income, number of employees, or total assets. Many studies have confirmed the validity of Zipf's law for firm sizes existing at current time t and estimated with these different measures (Simon and Bonini, 1958; Ijri and Simon, 1977; Sutton, 1997; Axtell, 2001; Okuyama et al., 1999; Gaffeo et al., 2003; Aoyama et al., 2004; Fujiwara et al., 2004a,b; Takayasu et al., 2008).

Initially formulated as a rank-frequency relationship quantifying the relative commonness of words in natural languages (Zipf, 1949), Zipf himself recognized in his book the general relevance to this law to the distribution of city sizes, among others. Many works have since shown that Zipf's law indeed accounts well for the distribution of city sizes (see for a review Gabaix, 1999 and references therein), as well as firm sizes all over the world, as just mentioned. Zipf's law has also been found in Web access statistics and Internet traffic characteristics (Glassman, 1994; Nielsen, 1997; Adamic and Huberman, 2000; Barabasi and Albert, 2002; and with deviations Breslau et al., 1999), in inbound degree distributions over Web pages (Kong et al., 2008), in weekend gross per theater for a movie scaled by the average weekend gross over its theatrical lifespan (Sinha and Pan, 2006), in bibliometrics, informetrics, scientometrics, and library science (Adamic and Huberman, 2002, and references therein) and in the distribution of incoming links to packages found in different Linux open source software releases (Maillart et al., 2008). Sinha and Pan (2006) provides a rather exhaustive review of the many power laws found in the distribution of human activities. There are also suggestions for applications to other physical and biological, sociological and

A. Saichev et al., *Theory of Zipf's Law and Beyond*,
Lecture Notes in Economics and Mathematical Systems 632,
DOI 10.1007/978-3-642-02946-2_1, © Springer-Verlag Berlin Heidelberg 2010

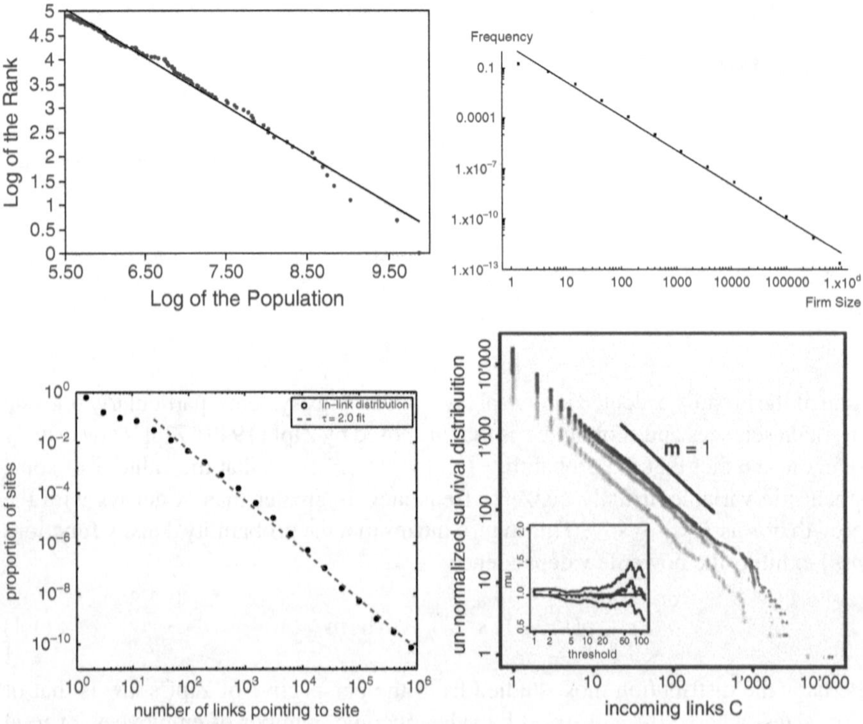

Fig. 1.1 Illustration of Zipf's law for city sizes (*upper left panel*, reproduced from Ioannides and Gabaix, 2003), for firm sizes (*upper right panel*, reproduced from Axtell, 2001), for the number of Internet links pointing to some website (*lower left panel*, reproduced from Adamic and Huberman, 2002) and for the number of incoming links to packages found in different Linux open source software releases (*lower right panel*, reproduced from Maillart et al., 2008)

financial market processes. For instance, using data from gene expression databases on various organisms and tissues, including yeast, nematodes, human normal and cancer tissues, and embryonic stem cells, Furusawa and Kaneko (2003) found that the abundances of expressed genes obey Zipf's law. See the list of references in http://linkage.rockefeller.edu/wli/zipf/index_ru.html. Figure 1.1 illustrates several cases where Zipf's law holds for different fields of social and natural sciences.

We should point out that there are some dissenting notes. For instance, several works have suggested that, for the distribution of firm's sizes, the log-normal distribution may actually be a better model than Zipf's law (Stanley et al., 1995; Cabral and Mata, 2003; Kaizoji et al., 2006; Duchin and Levy, 2008; Schwarzkopf and Farmer, 2008). The issue is confounding because often the authors are not always speaking of the same thing. Stanley et al. (1995)'s result has now been understood as due to an incomplete database, missing most of the small firms and hence biasing the distribution downward towards the log-normal shape for small firms (Axtell, 2001). Axtell (2001) has shown that firm' sizes measured by the number of employees, by the total sales or by the economic capital (debt + equity) are all consistently obeying Zipf's law. From an economic view point, it can indeed be expected that these three firm characteristics are globally proportional to each other in a same industry branch, or for a same business model, so that if Zipf's law holds for one of them, it should hold for the others. On the other hand, equity provides only a part of the economic capital of a firm, which depends on the financing strategies chosen by the firm, in addition to the impact of the stock market fluctuations. It is not clear that the financing strategies are stationary as a function of time, except perhaps for mature firms with no more any innovation or M&A (mergers and acquisitions) for which the financial structure of the firm (its debt/equity ratio) may be approximately constant. Therefore, the fact that Zipf's law may not be the best model for the distribution of equity sizes (Duchin and Levy, 2008) is not surprising. Another issue is the possible slow convergence of the distribution to its expected asymptotic long-time shape (Schwarzkopf and Farmer, 2008). Difference between countries due to the presence of specific financial constraints may be also an issue (Cabral and Mata, 2003).

Kitov (2009) points out that the significant differences in the evolution of firm size distribution for various industries in the United States puts important constraints on the modelling of firm growth. This line of thought opens the road toward linking asset pricing models, investment strategies and firm growth processes. In this spirit, Malevergne and Sornette (2007) have discovered a new endogenous pricing factor resulting from the heavy-tailed distribution of firm sizes, which has empirically a similar explanatory power as the phenomenological Fama–French three-factor model (Fama and French, 1993, 1995).

Employing Census 2000 data to create the most extensive and thorough investigation to date of the distribution of city sizes in the USA, Eeckhout (2004) reported that the empirical distribution follows a log-normal distribution rather than Zipf's law. Reanalysing this data, Levy (2009) confirms that the log-normal distribution indeed provides an excellent fit to the empirical data for 99.4% of the size range. However, for the top 0.6% of largest cities, the empirical distribution is dramatically different from the log-normal, and follows a power law. Levy notes that, while this top part of the distribution involves only 0.6% of the cities, it is extremely important as it accounts for more than 30% of the sample population. This type of hybrid log-normal-power-law distribution will find a natural explanation in the framework that we develop in the following chapters, and in particular in Chap. 6. The debate is however not closed as Eeckhout (2009) argues that the deviations from the log-normal model identified visually by Levy (2009) can be expected from the confidence bands generated by the Lilliefors test with 5% significance level. The problem however is that Eeckhout (2009)'s argument is based on a very weak test: the Lilliefors test, an adaptation of Kolmogorov–Smirnov test, is inadequate to identify deviations that occur in the tail, since its statistics is constructed from the maximum discrepancy between the log-normal and the empirical distribution. Anderson–Darling tests, for instance, are more adapted to the problem of distinguishing distributions in their tails (Malevergne et al., 2005; Malevergne and Sornette, 2006). In a forthcoming paper, Malevergne et al. (2009) develop a more powerful test specifically designed to compare the log-normal family to the power law family, which confirms quantitatively the intuition of Levy (2009). In order to address the issues associated with the definition of a city (administrative or geographic), Rozenfeld et al. (2009) employ a recently proposed clustering algorithm Rozenfeld et al. (2008) to construct cities from the bottom-up, without administrative data, but by using geographical proximity. They find that Zipf's law holds for cities above 10,000 inhabitants in the USA, and above 1,000 inhabitants in the UK.

Among the many more or less successful explanations proposed to understand the origin of Zipf's law, one of the most promising seems to be the explanation by Gabaix (1999) and Ioannides and Gabaix (2003) formulated in the context of the distribution of city sizes, based on Gibrat's law. Gabaix (1999) assumed that each city exhibits a stochastic growth rate distributed independently from its present size. Gibrat's law for city growth (together with some deviations of Gibrat's law for small sizes), normalized to the whole population of a given country, then leads to distributions of city sizes very close to Zipf's law. In general terms, Gibrat's law amounts to assume a stochastic multiplicative process. Such processes are found in many economic as well as natural systems (Sornette, 2006, and references therein). As a recent illustration, Clauset and Erwin (2006) explain in this way (with the inclusion of a mechanism involving size-dependent long-term extinction risks) the evolution and distribution of biological species body sizes.

However, the derivation of Zipf's law from the pure Gibrat's rule suffer from a few problems. First, the exact scale-independent Gibrat's law leads to a log-normal distribution of sizes, which is not a power law and only slowly approaches to a power law in the limit of large log-variance (and some other conditions), becoming at the same time more and more degenerate. Some additional assumptions are therefore needed in order to produce the stable non-degenerate Zipf's law. In particular, Gabaix (1999) assumed that, for cities of small sizes, there are some exogenous factors preventing further decaying of their population (see also Levy and Solomon, 1996; Malcai et al., 1999). More appropriate to social and economic phenomena is to allow for eliminating cities or firms as they reach a small size. An example is the transition from city to rank of village as the size goes below some threshold.

More generally, it is important to take into account the continuous process with births and deaths, which plays a central role at time scales as short as a few years. This is in contrast with Gabaix's approach for instance based on the supposition, simplifying considerably the theoretical modeling, that all cities originate at the same instant t_0, and then only grow stochastically, obeying the balanced Gibrat's law mentioned above. This supposition is clearly falsified by empirical evidence, as discussed later in the book.

A goal of this book is to demonstrate that birth as well as death processes are especially important to understand the economic foundation of Zipf's law and its robustness. Yamasaki et al. (2006) have shown that a model of proportional growth of the existing firms in the presence of a steady influx of new firms leads to Zipf's law truncated by an exponential taper, without the need to modify Gibrat's law for small sizes. The exponential cut-off results from the finite life of the economy. Our general analysis encompasses these results and put them in a larger perspective. Expressed in the context of an economy of firms, we will consider two different mechanisms for the exit of a firm: (1) when the firm total asset value becomes smaller than a given minimum threshold (which can vary with time and with countries) and (2) when an exogenous shock occurs, modeling for instance operational risks, independently of the size of the firm. Of course, these two mechanisms have their counterparts in the different fields of application where Zipf's law is discussed.

The following chapters are built on the realistic description of the behavior of the asset value of firms (which is more dynamic than the formation of cities), according to which the births of firms occur according to a random point process characterized by some intensity $\nu(t)$. Jointly, one should take into account the well-documented evidence that firms die, for instance when their size goes under some low asset value level. It turns out that taking into account the random flow of firm births and deaths, in combination with Gibrat's law, leads to the pure and non-degenerate Zipf's law under a balance condition, without the need for the rather artificial modification of Zipf's law for small sizes. (We note that the fact that deviation of Gibrat's law has been documented for small firms is another issue, as the documented deviations do not necessarily obey the assumptions needed in Gabaix's derivation.) As a bonus, the approach in terms of the dynamics of birth–death together with stochastic growth, that we develop here, leads to specific predictions of the conditions under which deviations from Zipf's law occur, which help rationalize the empirical evidence documented in the literature. The conditions involve either deviations from Gibrat's law in the stochastic growth process of firms or the existence of an unbalanced growth or decay of the birth intensity $\nu(t)$ of new firms, as we explain in details below.

In the theory developed in the following chapters, we also take into account that the intensity of firm's births may increase exponentially, that the sizes of entrant firms and the minimum viable size may grow exponentially with time with additional random fluctuations, hence generalizing Blank and Solomon (2000). Putting all elements and results of our analyses together, we conclude that the explanation for the generic empirical evidence that the exponent m is close to 1 (Zipf's law) is likely due to the weak dependence of m on the different parameters of the problem. This renders unnecessary the question for why the parameters would combine to obey exactly the balance condition. The closeness of the exponent m to 1 for a large range of parameters is quantified for instance in Figs. 7.5 and 8.1.

For transparency of derivations and for convenience of analytic calculations, we use a continuous version of Gibrat's law, allowing us to benefit from the properties of the Wiener process and the mathematical framework of Kolmogorov's diffusion equations. We unearth new properties associated with the stochastic behavior of firm assets. We show that the death of firms at some low value level as well as the possibility of significant deviations from Gibrat's law do not affect the asymptotic validity of Zipf's law in the limit of large firm sizes. By analyzing a large class of diffusion processes modeling the behavior of firm assets with growth rates very different from Gibrat's condition, we find general conditions for the validity of Zipf's law. Specifically, we have discovered stochastic growth models with non-Gibrat properties, leading to Zipf's and related power laws for the current density of firm's asset values.

Our book does not cover the more economically based theories, in the spirit for instance of Lucas (1978), which developed a theory of size distribution of business firms based on an underlying distribution of managerial talents and the competitive process of allocation of productive factors. Similarly, we do not expand on the general equilibrium model of the distribution of firm sizes proposed by Luttmer (2007), in terms of primitives such as entry and fixed costs, and the ease with which entrant firms can imitate incumbent firms. Let us also mention Rossi-Hansberg and Wright (2007a) which develops a general equilibrium theory of economic growth in an urban environment. In this theory, variation in the urban structure through the growth, birth, and death of cities is the margin that eliminates local increasing returns to yield constant returns to scale in the aggregate. They show that scale-independent growth for a finite number of industries, combined with an empirically-based form of entry and exit and a lower bound for establishment sizes that converges to zero, is sufficient to generate an invariant distribution that satisfies Zipf's law. Rossi-Hansberg and Wright (2007b) present a theory of the establishment size dynamics based on the accumulation of industry-specific human capital that simultaneously rationalizes the economy-wide facts on establishment growth rates, exit rates, and size distributions. Using a simple model of market share dynamics with bounded rational consumers and firms interacting with each other, Yanagita and Onozaki (2008) find that, in an oligopolistic phase associated with intermediate greediness of agents, the market-share distribution of firms follows Zipf's law and the growth-rate distribution of firms follows Gibrat's law.

The book is organized as follows. Chapter 2 presents the continuous version of Gibrat's law and some peculiarities of the stochastic behavior of the geometric Brownian motion of firm's asset values, resulting from Gibrat's law.

Chapter 3 describes the proposed model for the current density of firm's asset values, taking into account the random flow of the birth of firms. We show that, if some natural balance condition holds, while the intensity of firms is independent of time ($\nu =$ const.), then the exact Zipf's law holds true.

Amazingly, despite the relevance of Gibrat's law and the corresponding geometric Brownian motion in a wide range of physical, biological, sociological and other applications, many researchers do not make use of many of the interesting properties exhibited by realizations of the geometric Brownian motion, in order to derive detailed explanations of Zipf's and related power laws. Thus, in Chap. 4, we gather useful properties of realizations of the geometric Brownian motion, which play a significant role for the understanding of the roots and conditions of the validity of Zipf's law.

Chapter 5 discusses in detail the influence on the validity of Zipf's law of the occurrence of the death of firms when their value falls below some low level. In Chap. 6, we derive an equation for the steady-state density of firm asset values, which enables us to explore in detail the consequences of deviations from Gibrat's law at moderate asset values on the validity of Zipf's law at higher asset values.

Chapter 7 is devoted to discussing the conditions for the existence of Zipf's law and the circumstances under which deviations from it occur, when taking into account the possibility for sudden death of firms occurring even for large sizes.

Chapter 8 provides the most general treatment taking into account time dependence of birth rates, sizes at birth, and minimum firm sizes. Chapters 7 and 8 show that, with all these additional ingredients, Zipf's law holds if a generalized balance condition is valid. In particular, we discuss the robustness of Zipf's law to variations of the mean birth rate and of the rate of growth of the mean asset value of particular firms. Moreover, we find that Zipf's law is "attracting" the power laws found in absence of the strict validity of the balance condition: as the volatility of the growth of firms increases, the power law distribution of firm's sizes becomes closer to Zipf's law, and the later is obtained asymptotically for very large volatilities for all values of the other parameters.

All previous chapters have emphasized the dynamics of the statistical average of various firm properties in the limit where the number of firms in the economy grows without bounds. Chapter 9 asks if the results described in previous chapters can be used for the description of a single realization of a finite economy, an issue of great importance for the application of our theory to empirical data. For this, we derive the statistics of the number of firms, the fluctuation characteristics of the size of the global economy and the size of fluctuations decorating the asymptotic Zipf's law for finite economies. We provide a simple estimation of the expected statistical deviations from Zipf's law and its range of validity for realistic parameters. This provides a benchmark for assessing the range of validity of Zipf's law in empirical data.

Chapter 10 concludes first by stressing the importance of the balance conditions for Zipf's law to hold. Then, we provide the nucleus of what could be a more complete mathematical theory of firm sizes, based on taking into account in addition the mergers between firms as well as it symmetric, the creation of spin-off firms from parents which outsource a part of their existing business as separate units. These economic events can be modeled by using the mathematics of coagulation-fragmentation processes, which are briefly described here in the context of the dynamics of firms. We provide only a preliminary introduction to encourage future works to tackle these complex and rich issues.

For clarity, consistence of language and conciseness, we will discuss the origin and conditions of the validity of Zipf's law using the terminology of financial markets and referring to the density of the firm's asset values. We use firms at the entities whose size distributions are to be explained. It should be noted, however, that most of the relations discussed in this book, especially the intimate connection between Zipf's and Gibrat's laws, underlie Zipf's law in diverse scientific areas. The same models and variations thereof can be straightforwardly applied to any of the other domains of application.

Chapter 2
Continuous Gibrat's Law and Gabaix's Derivation of Zipf's Law

In this chapter, we describe in detail the continuous version of Gibrat's law and explain its close connection with the geometric Brownian motion (GBM), underlying any scale independent stochastic process. Due to the importance of the GBM for many economical, physical, biological and sociological applications, we focus our attention on the basic key properties of GBM. Some more subtle statistical properties of the GBM necessary for a deep understanding of the behavior of its realizations and, ultimately, the corresponding power distributions, are discussed in the following chapters. Although the GBM adequately simulates stochastic processes occurring in various scientific fields, here and for the remaining of the book, we use the terminology of firm's asset values.

2.1 Definition of Continuous Gibrat's Law

Let $S(t)$ be the current asset value of some given firm, established at the instant $t = 0$. The corresponding growth rate of the firm's asset value within the time window $[t, t + \Delta]$ is equal, by definition, to the ratio

$$R(t, \Delta) := \frac{S(t + \Delta) - S(t)}{S(t)}, \qquad t \geq 0, \quad \Delta > 0. \tag{2.1}$$

According to Gibrat's law, the growth rate $R(t, \Delta)$ of the firm's asset value is independent of the asset value $S(t)$. Thus, if a given firm was born at the instant $t = 0$, then at the time $t = n\Delta$, where n is an arbitrary positive integer, one may represent the firm's asset value as the product of n independent factors

$$S(n\Delta) = s_0 \prod_{k=0}^{n-1} \left[1 + R(k\Delta, \Delta)\right], \tag{2.2}$$

where $s_0 > 0$ is the firm's initial asset value. Correspondingly, the logarithm of the current asset value is equal to the sum of the statistically independent summands

A. Saichev et al., *Theory of Zipf's Law and Beyond*,
Lecture Notes in Economics and Mathematical Systems 632,
DOI 10.1007/978-3-642-02946-2_2, © Springer-Verlag Berlin Heidelberg 2010

$$Y(n\Delta) := \ln S(n\Delta) - \ln s_0 = \sum_{k=0}^{n-1} r(k\Delta), \tag{2.3}$$

where

$$r(k\Delta) := \ln\left[1 + R(k\Delta, \Delta)\right]. \tag{2.4}$$

In what follows, we will suppose, mostly for simplicity, that the sequence $\{r(k\Delta)\}_{k\in\mathbb{N}}$ is statistically homogeneous in the sense that $\{r(k\Delta)\}_{k\in\mathbb{N}}$ are identically independently distributed (i.i.d.) random variables, and such that their mean and variance are finite

$$\mathrm{E}[r(\Delta)] < \infty, \qquad \mathrm{E}[r(\Delta)^2] < \infty. \tag{2.5}$$

There are two interpretations of the above relations, the purely mathematical and the "real life" ones. The former implies that elements of the sequence $\{r(k\Delta)\}_{k\in\mathbb{N}}$ are i.i.d. for any, even infinitesimal, time step Δ. This means in turn that, for any $t > 0$, $Y(t)$ is a Wiener process with drift, which can thus be written as

$$Y(t) := Y(t, c, b) = ct + bW(t), \tag{2.6}$$

where $W(t)$ is a standard Wiener process. We recall that

Definition 2.1.1. The stochastic process $\{W(t),\ t \geq 0\}$ is a standard Wiener process starting from 0 on the probability space $(\Omega, \mathcal{F}, \mathbb{P})$ if:

1. $\mathbb{P}(W(0) = 0) = 1$
2. $\forall\, 0 \leq s \leq t$, the random variable $W(t) - W(s)$ follows the normal distribution with mean 0 and variance $t - s$
3. $\forall\, 0 = t_0 < t_1 < \cdots < t_p$, the random variables $(W(t_k) - W(t_{k-1}), 1 \leq k \leq p)$ are independent

The coeffcients c and b in (2.6) respectively denote the drift and the intensity of the stochastic component of the behavior of the logarithm of the firm's asset value. They satisfy the relations

$$c\Delta = \mathrm{E}[Y(t+\Delta)] - \mathrm{E}[Y(t)], \quad b^2\Delta = \mathrm{Var}[Y(t+\Delta) - Y(t)]. \tag{2.7}$$

In the "real life" interpretation, relation (2.7) is valid only asymptotically for $\Delta \gtrsim \Delta_c$, where Δ_c is some characteristic time scale of the random behavior of the firm's asset value. In this case, equality (2.6) is only asymptotically true, due to the Central Limit Theorem, for large enough $t \gg \Delta_c$. We will suppose that t is large enough, so that the relation (2.6) holds true. Correspondingly, $Y(t, c, b)$ defined by (2.6) is a Wiener process with drift, whose pdf is Gaussian and given by

$$\phi(y;t) := \frac{1}{\sqrt{2\pi b^2 t}} \exp\left(-\frac{(y-ct)^2}{2b^2 t}\right). \tag{2.8}$$

2.2 Geometric Brownian Motion

It follows from the above reasoning that, in the framework of the continuous Gibrat's law, the logarithm of the firm's current asset value, normalized by the initial value, $S(t)/s_0$, is a Wiener process with drift (2.6), so that $S(t)$ is proportional to the exponential of the Wiener process with drift

$$S(t) = s_0 e^{ct+bW(t)}. \tag{2.9}$$

Note also that (2.9) explicitly expresses the scale invariance of the stochastic behavior of the firm's asset value. Namely

$$S(t) = s_0 X(t, c, b), \tag{2.10}$$

where the geometric Brownian motion (hereafter GBM)

$$X(t, c, b) := e^{Y(t,c,b)} = e^{ct+bW(t)} \tag{2.11}$$

does not depend on the initial asset value s_0. This scale invariance is a direct consequence of Gibrat's law of proportional growth.

An alternative and maybe more transparent explanation of the relation between Gibrat's law and the geometric Brownian motion can be obtained by considering the stochastic differential equation (SDE) solved by the GBM. The Wiener process with drift $Y(t)$ given by (2.6) is solution to

$$dY(t) = c\,dt + b\,dW(t). \tag{2.12}$$

Ito calculus allows us to derive the SDE solved by the GBM $S(t)$:

$$dS(t) = a \cdot S(t)\,dt + b \cdot S(t)\,dW(t), \tag{2.13}$$

where

$$a := c + \frac{b^2}{2} \tag{2.14}$$

denotes the *expected* growth rate of a single firm. It clearly appears that the growth rate

$$R(t, dt) = \frac{dS(t)}{S(t)} = a\,dt + b\,dW(t), \tag{2.15}$$

is independent of the current firm size, as stated by Gibrat's law.

2.3 Self-Similar Properties of the Geometric Brownian Motion

One can represent the GBM in a form convenient for analytic calculations

$$X(t, c, b) = \chi(t/t_b, \lambda), \tag{2.16}$$

where

$$\chi(\tau, \lambda) := e^{\lambda\tau + \omega(\tau)} \tag{2.17}$$

is the GBM expressed in terms of the dimensionless argument τ:

$$\tau := t/t_b, \qquad t_b := 2/b^2, \tag{2.18}$$

defining t_b as a characteristic time associated with the volatility of the firm asset values. Here, λ is also a dimensionless parameter

$$\lambda := ct_b = 2c/b^2, \tag{2.19}$$

while $\omega(\tau)$ is the Wiener process in the variable τ, such that

$$\mathrm{E}[\omega(\tau)] = 0, \qquad \mathrm{E}[\omega^2(\tau)] = 2\tau. \tag{2.20}$$

Equivalently, $\omega(\tau) \stackrel{law}{=} \sqrt{2}\, W(\tau)$. The Wiener process possesses the following self-similar (fractal) property

$$\xi\omega(\tau) \stackrel{law}{=} \omega(\xi^2\tau), \tag{2.21}$$

where, here and below, the symbol $\stackrel{law}{=}$ expresses the statistical equivalence, i.e., equality in law. Accordingly, the GBM $\chi(\tau, \lambda)$ obeys the following statistical equivalence relation

$$\chi(\tau, \lambda) \stackrel{law}{=} \chi^{1/\lambda}(\lambda^2\tau, 1). \tag{2.22}$$

This leads, in particular, to the following inverse relation, which is useful in what follows

$$\chi(\tau, \lambda) \stackrel{law}{=} \chi^{-1}(\tau, -\lambda) \quad \Longleftrightarrow \quad X(t, c, b) \stackrel{law}{=} X^{-1}(t, -c, b). \tag{2.23}$$

2.4 Time Reversible Geometric Brownian Motion

We will call the GBM $X(t, c, b)$ defined by (2.11) *time reversible* if the growth rate $R(t, \Delta)$ of asset values (2.1) possesses the same statistical properties both forwards and backwards in time. Specifically,

Definition 2.4.1. The GBM $S(t)$ is a time reversible GBM if the growth rate $R(t, \Delta)$ defined by (2.1) forwards in time is statistically equivalent to the growth rate backwards in time, defined by

$$R_-(t,\Delta) := \frac{S(t) - S(t+\Delta)}{S(t+\Delta)}. \tag{2.24}$$

Obviously, a necessary and sufficient condition of time reversibility of the diffusion process $S(t)$ given by (2.10)–(2.11) is

$$c = 0 \quad \Longleftrightarrow \quad \lambda = 0. \tag{2.25}$$

This condition separates the whole family of GBM diffusion processes into two qualitatively different sub-families. Namely, there are stochastically growing ($c > 0$) and stochastically decaying ($c < 0$) GBMs. The precise sense of the above terminology will become clearer when we will discuss in detail the probabilistic properties of realizations of GBM (see Chap. 4). Here we note only that the two sub-families generate qualitatively different stochastic phenomena[1] which, nevertheless, give both birth to power law distributions.

2.5 Balance Condition

In Gabaix (1999)'s explanation of Zipf's law, in addition to Gibrat's law, a normalization condition plays a crucial role. We reformulate it in a somewhat different form, appropriate for our following derivations. Let there be N firms at the initial instant $t = 0$, whose asset values are equal to s_{0i}, $i = 1, \dots, N$, so that the initial overall asset value Υ_0 reads

$$\Upsilon_0 = \sum_{i=1}^{N} s_{0i}. \tag{2.26}$$

Suppose that the stochastic behavior of each firm's asset value obeys the continuous Gibrat's law, leading to the following expression of the size of firm i [according to (2.10)]

$$S_i(t) = s_{0i} X_i(t), \tag{2.27}$$

where $\{X_i(t)\}$ are statistically equivalent GBMs, which may be but do not need to be independent. Correspondingly, the current mean overall asset value is the sum of the mean asset value of each firm

$$\Upsilon(t) = \sum_{i=1}^{N} \mathrm{E}[S_i(t)] = \Upsilon_0 \mathrm{E}[X(t,c,b)], \tag{2.28}$$

[1] For example $c < 0$ is inherent to the stochastic behavior of firm's asset values, while $c > 0$ is typical of a number of genera and their growth, as described by models of biological evolution.

where $\mathrm{E}[X(t,c,b)]$ is the mean of the GBM, simulating the stochastic behavior of the asset value of any firm. It follows from the statistical properties of the Wiener process that the moment of order n of the GBM is equal to

$$\mathrm{E}[X^n(t)] = \exp\left(nct + \frac{n^2 b^2}{2} t\right). \tag{2.29}$$

Thus, the mean of the current overall asset value is

$$\Upsilon(t) = \Upsilon_0 e^{at}. \tag{2.30}$$

It justifies the denomination of *expected growth rate* given to the coefficient a defined by (2.14).

Gabaix (1999) introduced the normalization condition, which demands that the mean of the overall asset value does not depend on time. At face value, this condition describes a stationary economy. Because real economies grow (at least at times without significant wars or crises in the western world), this condition needs to be generalized with a change of frame growing with the global economy.

Let us first discuss the simple case where the economy is stationary. We will call this condition the *balance condition*. Mathematically, it means that

Definition 2.5.1. In Gabaix (1999)'s model, the economy follows a *balance growth path* if its expected growth rate equals zero: $a = 0$.

This condition is obviously equivalent to $c = -\frac{b^2}{2}$, or $\lambda = -1$ [where λ is defined in (2.19)] and ensures that

$$\Upsilon(t) = \Upsilon_0 = \text{const.} \tag{2.31}$$

One can explain the balance condition as a consequence of some kind of equal opportunities (in a mean sense) for all firms established at the same time t_0 (above, we have chosen for convenience of analysis $t_0 = 0$).

Note that the balanced GBM, in the sense of Definition 2.5.1, is stochastically decaying since $c < 0$. Roughly speaking, due to the stochastic decay of the balanced asset value $S(t)$, after some time only a few firms among those established simultaneously still keep a non-vanishingly small size. We will return to discussing this dramatic peculiarity of firm's size under the balance condition first in Sect. 2.7 and then in more details in Chap. 5. It is worthwhile emphasizing additionally that, in some unstable growing or slumping economic era, where the overall asset value might exponentially grow, decay or change periodically, the balance condition might be violated.

2.6 Log-Normal Distribution

Similarly to the purpose of Gabaix (1999)'s paper, we intend to derive Zipf's law based on the continuous Gibrat's law and the balance condition. Firstly, in order to better reveal the hidden assumptions underlying Zipf's law, we point out some statistical properties of the GBM defined by (2.11) and describe the stochastic behavior of some given firm's asset value. Simple analytical calculations show that the pdf of the GBM $X(t, c, b)$ is log-normal

$$\psi(x;\tau) := \frac{1}{2\sqrt{\pi\tau}x}\exp\left(-\frac{(\ln x - \lambda\tau)^2}{4\tau}\right). \tag{2.32}$$

The log-normal distribution can be shown to be asymptotic to a power law for intermediate values of x in the following sense. We rewrite equality (2.32) in the form

$$\psi(x;\tau) = \psi_0(\tau)\, x^{-1-\zeta(x,\tau)}, \tag{2.33}$$

where

$$\psi_0(\tau) := \frac{1}{2\sqrt{\pi\tau}}\exp\left(-\frac{\lambda^2}{4}\tau\right) \tag{2.34}$$

and

$$\zeta(x,\tau) := \frac{\ln x}{4\tau} - \frac{\lambda}{2}. \tag{2.35}$$

If the second term of the r.h.s. of (2.35) prevails over the first term, i.e., if

$$\tau \gg \tau_p(x), \qquad \tau_p(x) := \frac{1}{2}\left|\frac{\ln x}{\lambda}\right|, \tag{2.36}$$

then one can neglect the first term and replace the log-normal distribution (2.32) by the power law

$$\psi(x;\tau) \simeq \psi_0(\tau)\, x^{\lambda/2-1}. \tag{2.37}$$

In particular, if the balance condition is valid (i.e., $\lambda = -1$), then we have

$$\psi(x;\tau) \sim x^{-3/2}. \tag{2.38}$$

Thus, when the balance condition holds for the GBM $X(t, c, b)$, the distribution of firm sizes is very well approximated at long times by a power law distribution, but the exponent $m = 1/2$ is different from the value $m = 1$ required for Zipf's law (1.1) to hold. Figure 2.1 shows the log-log plot of the log-normal distribution (2.32), illustrating the existence of the asymptotic power law (2.38) for intermediate sizes x's. As τ increases, the inequality in (2.36) is better and better verified and Fig. 2.1 indeed shows that the log-normal distribution (2.32) is better and better approximated by the power law (2.38). We stress in particular the very large range (number of decades) over which this asymptotic correspondence can hold.

Expression (2.37) shows that Zipf's law is obtained in this framework for the special unbalanced case $\lambda = -2$ for which the log-normal distribution (2.32) does have Zipf's law asymptotics $\psi \sim x^{-2}$. This is one of the several mechanisms found in the present book for Zipf's law. We do not emphasize it because it requires a special tuning of the parameter λ to a value which is not economically realistic insofar as it yields an exponentially decaying overall asset value: $\Upsilon(t) = \Upsilon_0 e^{-\frac{b^2}{2}t}$.

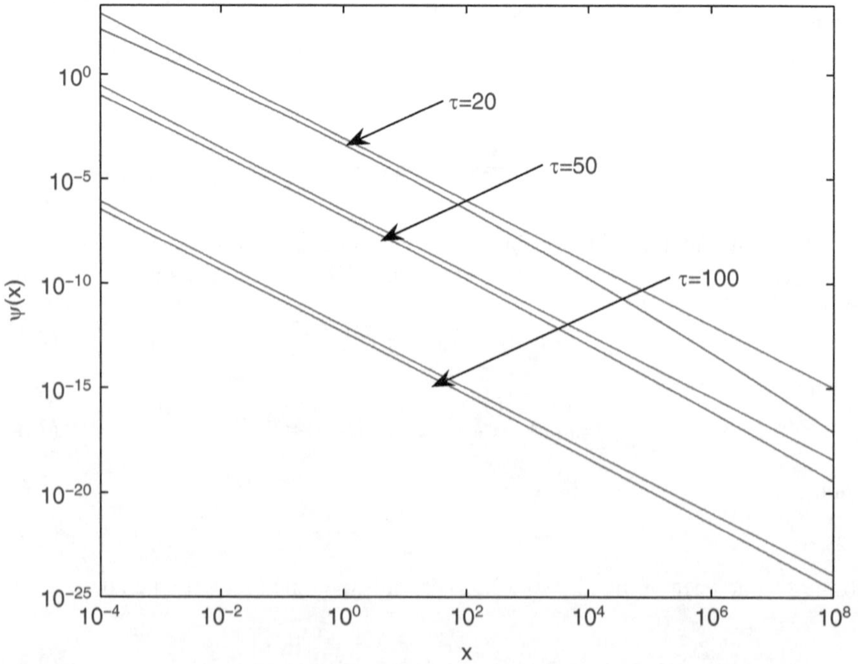

Fig. 2.1 Log-log plot of the log-normal distribution (2.32). *From top to bottom* $\tau = 20, 50, 100$. The corresponding *straight lines* indicate a pure power law $\sim x^{-3/2}$

2.7 Gabaix's Steady-State Distribution

The previous subsection has shown that Gibrat's law and the balance condition are not sufficient to obtain Zipf's law in general. Thus, in order to get Zipf's law, one needs some additional key supposition. One possible assumption was suggested by Gabaix (1999), which involves breaking Gibrat's law below some small threshold $s_{\min}$ in order to prevent cities/firms from becoming too small.

To illustrate more clearly the spirit of Gabaix's mechanism to prevent cities/firms from becoming too small, we provide a detailed mathematical explanation. Let $S(t)$ be some arbitrary diffusion process, whose pdf $f(s;t)$ is the solution to the diffusion equation

$$\frac{\partial f(s;t)}{\partial t} + \frac{\partial [a(s) f(s;t)]}{\partial s} = \frac{1}{2} \frac{\partial^2 [b^2(s) f(s;t)]}{\partial s^2} \tag{2.39}$$

satisfying some initial condition that reflects the statistical properties of the diffusion process $S(t)$ at the initial instant $t = 0$. If the value of $S(t)$ at $t = 0$ is known and is equal to s_0, then the initial condition takes the singular form

$$f(s;0) = \delta(s - s_0), \tag{2.40}$$

where $\delta(s - s_0)$ denotes Dirac's unit mass at s_0.

In the case of GBMs (2.10), we have

$$a(s) = a \cdot s\,, \qquad b(s) = b \cdot s, \tag{2.41}$$

where the factor a is defined by equality (2.14). The linear dependence of $a(s)$ and $b(s)$ as a function of the size s is nothing but a rephrasing of Gibrat's law of proportional growth. If the balance condition $a = 0$ holds, then (2.39) reduces to the form

$$\frac{\partial f(s;t)}{\partial t} = \frac{b^2}{2} \frac{\partial^2 [s^2 f(s;t)]}{\partial s^2}. \tag{2.42}$$

Using arguments similar to those which led us to the power laws (2.37) and (2.38), here gives analogously to (2.38)

$$f(s;t) \sim s^{-3/2} \quad \text{at large times.} \tag{2.43}$$

Accordingly, as already mentioned, to obtain Zipf's law for the balanced case, Gabaix was forced to break Gibrat's law at some small level $s_{\min} > 0$, to prevent cities from becoming too small. We will not reproduce here the details of Gabaix's specification on how cities are prevented from becoming too small. Instead, we propose another more mathematically convenient procedure, which reproduces Gabaix's results that Zipf's law is recovered asymptotically for large sizes s. Our specification consists in replacing the coefficient $b(s) = b \cdot s$ in (2.42) by

$$b(s) = b \cdot \sqrt{s^2 + s_{\min}^2}, \qquad s_{\min} > 0, \tag{2.44}$$

so that (2.42) is replaced by

$$\frac{\partial f(s;t)}{\partial t} = \frac{b^2}{2} \frac{\partial^2 [(s^2 + s_{\min}^2) f(s;t)]}{\partial s^2}, \tag{2.45}$$

and we supplement it with the reflection condition

$$\left. \frac{\partial f(s;t)}{\partial s} \right|_{s=0} = 0, \qquad t > 0. \tag{2.46}$$

The characteristic scale $s_{\min}$ violates Gibrat's law, which remains valid only asymptotically for $s \gg s_{\min}$ for which expression (2.44) reduces to the second equation in (2.41). Expression (2.44) and condition (2.46) embody two different mechanisms. The former expresses that the random relative changes of sizes of small cities/firms become more intensive than predicted by Gibrat's law. The later is the simplest device to prevent firms from dying, in the sense that their value is assumed to remain always positive. Together, these two expressions prevent cities/firms from becoming too small. Changing the reflection condition (2.46) into an absorption condition $f(s;t)_{s=0} = 0$ would lead to a fast decay of the number of firms and would prevent Zipf's law from appearing.

Then, an obvious non-degenerate normalized steady-state solution to (2.45) reads

$$f(s) = \frac{2}{\pi} \frac{s_{\min}}{s^2 + s_{\min}^2}, \qquad s > 0, \tag{2.47}$$

which recovers Zipf's law for $s \gg s_{\min}$. This concludes the simple presentation of the main idea in Gabaix (1999)'s paper to explain Zipf's law. An important point made clear by our derivation is that the specific way with which Gibrat's law is broken for small sizes s is not important as long as it conserves the key characteristics of preventing the small cities/firms from disappearing.

There is another, to our mind more natural, mechanism leading to Zipf's law, which does not require violating Gibrat's law at small firm value levels. It consists in taking into account the existence of incessant birth and death events among the population of firms occurring at successive random instants of time. As we show below, it turns out that firm birth and death processes are capable of creating power laws, and in particular Zipf's law in the balanced case $\lambda = -1$, without any violation of Gibrat's law. Moreover, these power laws are robust for even significant deviations from Gibrat's law. The following chapters are devoted to presenting detailed discussions of the connection between the process of birth and death of firms and the power laws for the mean density of firm's asset values.

Chapter 3
Flow of Firm Creation

3.1 Empirical Evidence and Previous Works on the Arrival of New Firms

The failure of the approach based solely on Gibrat's principle stems, at least in part, from the fact that it attempts to derive the distribution of firm sizes directly from the distribution of the asset value of a single firm. Indeed, many models start with the implicit or explicit assumption that the set of firms under consideration was born at the same origin of time. This approach is equivalent to considering that the economy is made of only one firm. Therefore, the distribution of firm sizes can reach a steady-state if and only if the distribution of the asset value of a single firm reaches a steady state, which seems rather counterfactual.

An alternative approach to model a stationary distribution of firm sizes is to account for the fact that firms do not appear at the same time but are born according to a more or less regular flow of newly created firms. For instance, Bonaccorsi Di Patti and Dell'Ariccia (2004) report a yearly average (over all industry branches) rate of birth for the period 1996–1998 equal to 5.6% for Italian firms with a maximum of 32% in some industry branches. Reynolds et al. (1994) give the regional average firm birth rates (annual firm births per 100 firms) of several advanced countries in different time periods: 10.4% (France; 1981–1991), 8.6% (Germany; 1986), 9.3% (Italy; 1987–1991), 14.3% (United Kingdom; 1980–1990), 15.7% (Sweden; 1985–1990), 6.9% (United States; 1986–1988). They also document a large variability from one industrial sector to another.

Simon (1955) was the first to address the question of the arrival of new firms (see also Ijri and Simon, 1977). He proposed to modify Gibrat's model by accounting for the entry of new firms over time as the overall industry grows. In Simon's model, at each time step, a new investment opportunity is created. It is either captured by a new entrant (probability p) that settles a new firm with unit size or by an already running firm (probability $(1 - p)$) whose size grows by one unit. In the later case, the probability that an existing firm seizes the next investment opportunity is proportional to its current size, in accordance with Gibrat's rule. Simon showed that

A. Saichev et al., *Theory of Zipf's Law and Beyond*,
Lecture Notes in Economics and Mathematical Systems 632,
DOI 10.1007/978-3-642-02946-2_3,

this model leads to a steady-state distribution of firm sizes with a regularly varying upper tail whose exponent is given by $m = 1/(1-p)$.[1]

The drawback of Simon (1955)'s approach and of subsequent related developments comes from the fact that the growth of existing firms is assumed to be strongly coupled with the entry of new firms. Either an existing firm is allowed to grow at the exclusion of the creation of a new firm, or a new firm is created but then there is no growth of any existing firm. In addition, no more than one existing firm can grow at any given time step. These different assumptions are obviously counterfactual. Besides, in order to get Zipf's law, the probability p that a new firm is created must be vanishingly small since $m = 1$ only in the limit where p goes to zero. Additionally, as noted by Krugman (1996), the convergence toward the steady-state is infinitely slow when p goes to zero.

In addition to Simon (1955)'s model, Gabaix (1999) accounts for the birth of new firms with the probability to create a new firm of size s being proportional to the current fraction of firms of that size and otherwise independent of time. However, this model does not reflect the real dynamics of firm's creation. For instance, using a data set from the Italian National Institute for Social Security dealing with 12 quarterly cohorts of new manufacturing firms (with at least one paid employee) born in each month of 1987, and their follow up until December 1992, Lotti and Santarelli (2004) find that the standard deviation of firm sizes is much smaller in the first quarter following their birth than at the end of the relevant period. Dispersion of firm sizes tends therefore to widen as surviving firms reach the minimum efficient scale level of output and specialize in one of the many clusters of products. Bartelsman et al. (2003) document that new entrant firms have a relatively small size compared with the more mature efficient size they develop as they grow. The effect is more pronounced in Canada and especially in the United States. In other words, it seems to be unrealistic to expect a non-zero probability for the birth of a firm of very large size, say, of size comparable to the largest capitalization currently in the market.[2]

Luttmer (2007) developed a general equilibrium model of the distribution of firm sizes, in terms of primitives such as entry and fixed costs, and the ease with which entrant firms can imitate incumbent firms. The tail index is a function of the population growth rate, the curvature parameter of the utility function of consumers, and of three technology parameters, which makes Zipf's law relying in this model on rather restrictive assumptions. The tail index however converges from above to the value 1 (Zipf's law) if either the technologies available to entrants improve at a rate that is only slightly above the rate at which the technologies of incumbents improve, or if the cost of entry becomes large relative to the fixed cost of operating a firm and as the extent to which new entrants lag behind incumbents in terms of productivity and market size becomes large. The later assumptions seem rather questionable.

---

[1] This model has recently been rediscovered under the name "preferential attachment" in the context of scale-free networks found in social networks, the world-wide web, or networks of proteins reacting with each other in the cell (Barabasi and Albert, 1999, 2002).

[2] We do not consider spin-off's or M&A (mergers and acquisitions).

3.2 Mathematical Formulation of the Flow of Firm's Births at Random Instants

A major qualitative difference between the approach developed here and Gabaix's approach is that he assumed that all cities (firms) were born at the same (possibly very ancient) time t_0. In contrast, we suppose that all firms present at the current time were born at random time instants $\{t_i\}$ in the past. A schematic illustration of this difference is shown in Fig. 3.1. This section describes the basic consequences of the existence of the random flow of firm's births at successive times $\{t_i\}$. For a more formal derivation, which relies on the theory of compound point processes, we refer the reader to Daley and Vere-Jones (2007). We will suppose for simplicity, but without loss of the main mechanism of firm creation, that the initial asset values of all firms at birth are the same and are equal to s_0. This assumption will be relaxed later. The value s_0 can be thought of as the minimal size below which a firm cannot be registered. In the case of cities, s_0 might be the minimal size of a

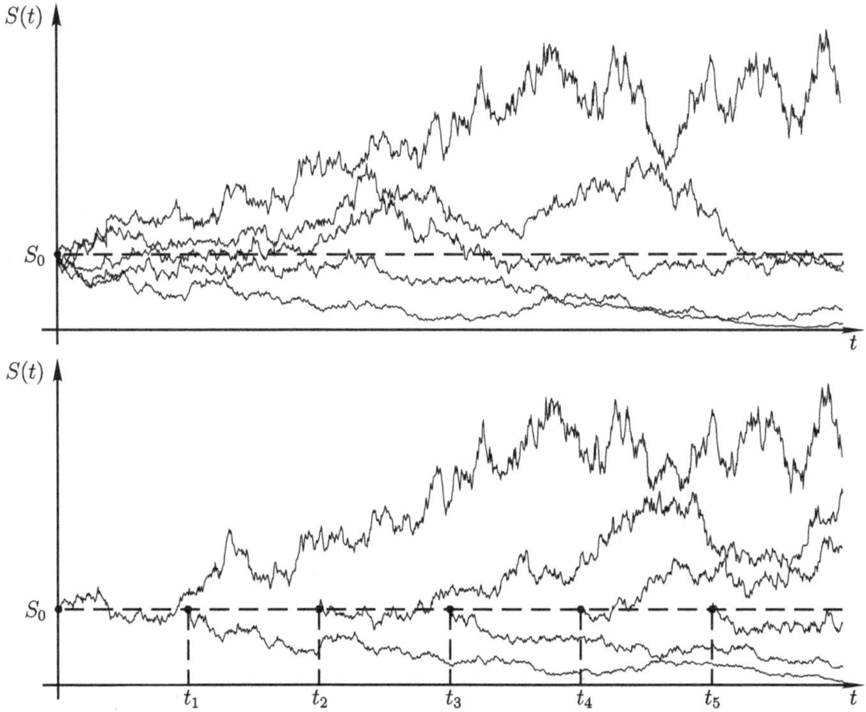

Fig. 3.1 A schematic illustration of Gabaix's and our models, leading to Zipf's law. Schematically depicted in the *upper plot* are graphs of sizes $S(t)$ of different firms (cities), which were all born at the same original time. The *lower plot* illustrates graphs of $S(t)$ of firms (cities) which were born at different instants. The *black filled circles* symbolize the instants of birth of subsequent firms (cities). The birth instants form in general a non-periodic time series

city, below which the settlement loses the status of city and becomes a village. As in the previous chapter, we will use, for consistency, the terminology of firm's asset values.

Definition 3.2.1. Let $\{t_i\}_{i \in \mathbb{N}}$ be an increasing random sequence representing the firm's birth dates

$$t_0 < t_1 < t_2 < \cdots < t_i < t_{i+1} \cdots, \tag{3.1}$$

where t_0 is the date of birth of the oldest firm. We call the point process $\{t_i\}_{i \in \mathbb{N}}$ the firm's *births flow*.

As a first pass, we do not make any additional assumption about the properties on the births flow. We will provide specific examples later on.

The second ingredient of our theory is the asset value $S_i(t, t_i)$, at time $t \geq t_i$, of the firm i born at time t_i.

Assumption 1 *Conditional on the date of birth t_i, the stochastic process that is the same for all firms, so that $S_i(t, t_i) \overset{law}{=} S(t, t_i)$ of the asset value of firm i.*

We are interested in the number $\tilde{G}(s, t)$ of firms whose asset values are smaller than s at time t. Obviously, $\tilde{G}(s, t)$ reads

$$\tilde{G}(s,t) := \sum_{i=0}^{\infty} \mathbf{1}\left[s - S(t, t_i)\right] \mathbf{1}(t - t_i), \tag{3.2}$$

where $\mathbf{1}(x)$ is the indicator function of the event $\{x \geqslant 0\}$. For any finite t, because there is an oldest firm, i.e., the economy started at some finite time in the past, this sum only involves a finite number of terms, and is therefore well-defined. We can rewrite the r.h.s. of (3.2) as the Lebesgue–Stieltjes integral

$$\tilde{G}(s,t) = \int_{t_0}^{t} \mathbf{1}\left[s - S(t, u)\right] d\tilde{N}(u), \tag{3.3}$$

where $\tilde{N}(u)$ is the counting measure, i.e., the largest integer i, satisfying the condition $t_i < u$.

We can now derive the expression of the expected value of $\tilde{G}(s, t)$ given by (3.3). We first consider the expected value of $\tilde{G}(s, t)$ conditional on all instants of the point process (3.1). It follows from (3.3) that this conditional expectation is equal to

$$\mathrm{E}\left[\tilde{G}(s,t)\Big| \{t_i\}\right] = \int_{t_0}^{t} \mathrm{E}\left[\mathbf{1}\left[s - S(t, u)\right] | \{t_i\}\right] d\tilde{N}(u). \tag{3.4}$$

Under the additional assumption

Assumption 2 *The stochastic process $S(t, u)$ are independent of the births flow $\{t_i\}_{i \in \mathbb{N}}$ for all u,* then

$$\mathrm{E}\left[\mathbf{1}\left[s - S(t,u)\right] | \{t_i\}\right] = \mathrm{E}\left[\mathbf{1}\left(s - S(t,u)\right)\right]. \tag{3.5}$$

Consequently, denoting by $F(s;t|u)$ the cumulative distribution function (cdf) of $S(t,u)$

$$F(s;t|u) := \Pr\{S(t,u) \leq s\} = \mathrm{E}\left[\mathbf{1}\left(s - S(t,u)\right)\right], \tag{3.6}$$

we can rewrite equality (3.4) in the form

$$\mathrm{E}\left[\tilde{G}(s,t)\Big| \{t_i\}\right] = \int_{t_0}^{t} F(s;t|u)\, d\tilde{N}(u). \tag{3.7}$$

By the law of iterated expectations, the mean number of firms whose size is less than s at time t

$$G(s,t) := \mathrm{E}\left[\tilde{G}(s,t)\right] \tag{3.8}$$

is equal to

$$G(s,t) = \mathrm{E}\left[\mathrm{E}\left[\tilde{G}(s,t)\Big| \{t_i\}\right]\right] = \int_{t_0}^{t} F(s;t|u) dN(u), \tag{3.9}$$

where

$$N(u) := \mathrm{E}\left[\tilde{N}(u)\right] \tag{3.10}$$

is the average number of firms born up to time u.

Everywhere below we assume, for simplicity, that

Assumption 3 *The average number of firms born up to time u, $N(u)$, is absolutely continuous with respect to Lebesgue measure so that it admits a density $\nu(u)$.*

This density, which will be called the *intensity of firms births*, allows us to state that

Proposition 3.2.1. *Under the Assumptions 1–3, the mean distribution of firm's size reads*

$$G(s,t) = \int_{t_0}^{t} \nu(u) F(s;t|u) du. \tag{3.11}$$

In what follows, it will be more convenient to deal with the derivative of $G(s,t)$ with respect to s

$$g(s,t) := \frac{\partial G(s,t)}{\partial s}. \tag{3.12}$$

We call this derivative the *mean density* of firm asset's values, which is defined for firms which are in business at the current time t. Differentiating both sides of equality (3.11) with respect to s, and assuming that $F(s;t|u)$ admits an integrable derivative with respect to s

$$f(s;t|u) := \frac{\partial F(s;t|u)}{\partial s}, \tag{3.13}$$

we obtain from Proposition 3.2.1 the following result

Corollary 3.2.1. *Under the assumptions of Proposition 3.2.1 and assuming additionally that the cumulative distribution $F(s,t|u)$ of $S(t,u)$ admits a integrable derivative with respect to s, the mean density of firm's size reads*

$$g(s,t) = \int_{t_0}^{t} \nu(u) f(s;t|u) du. \tag{3.14}$$

It relates the mean density of firms size $g(s,t)$ to the probability density function (pdf) $f(s;t|u)$ of the size s at time t of some particular firm born at time $u < t$.

In order to simplify one step further the result of Corollary 3.2.1, let us introduce the notion of homogeneous stochastic growth

Definition 3.2.2. The stochastic growth of firm asset values is *homogeneous* if $f(s;t|u) = f(s;t-u)$.

It allows us to state that

Corollary 3.2.2. *Under the assumption of Corollary 3.2.1, the mean density of firm's size for homogeneously stochastically growing firms reads*

$$g(s,t) = \int_{0}^{t-t_0} \nu(t-u) f(s;u) du. \tag{3.15}$$

Remark 3.2.1. Taking here for simplicity but without loss of generality the origin time of the economy as the time origin $t_0 = 0$, we finally get

$$g(s,t) = \int_{0}^{t} \nu(t-u) f(s;u) du. \tag{3.16}$$

The relations (3.14) and (3.16) hold for arbitrary stochastic births flows $\{t_i\}$, only assuming that (1) the stochastic processes $S(t,t_i)$ and the birth flow $\{t_i\}$ are independent, and (2) that there exists a continuous mean birth rate $\nu(t)$. To make explicit the hidden springs underlying the concept of a mean rate $\nu(t)$, we present in appendix some particular examples of statistics for typical births flow $\{t_i\}$.

3.3 Existence of a Steady-State Distribution of Firm's Sizes

We proved in the previous section that the mean density $g(s,t)$ is given by the relation (3.16) when we take into account the flow of firm's births, provided that the births flow $\{t_i\}$ in Definition 3.2.1 and the stochastic processes $S(t,t_i)$ of firms assets values are statistically independent. It turns out that, under some natural conditions, the mean density $g(s,t)$ admits, as t goes to infinity, a limit expression (independent of t):

Definition 3.3.1. If there exists a positive integrable function $g(s)$ such that

$$g(s) := \lim_{t\to\infty} g(s,t),$$

it is called the *steady-state mean density of firms size.*

In what follows, for the sake of pedagogy of the exposition, we will omit some details of the derivations and corresponding proofs for the general case. We however present in some depth the specific case where the mean rate of firm's birth is constant, so that (3.16) reads

$$g(s,t) = \nu \int_0^t f(s;u)du \tag{3.17}$$

and provide rigorous arguments and conditions, ensuring the existence of a steady-state density.

The existence of a steady-state mean density is equivalent to the convergence of the improper integral

$$g(s) = \nu \int_0^\infty f(s;u)du. \tag{3.18}$$

It is well-known that, for some given s, the improper integral converges, if the pdf $f(s;u)$, as a function of u, tends to zero, for $u \to \infty$, faster than the power function u^{-p}, where $p > 1$. In particular, the following condition of convergence is sufficient for the existence of the integral (3.18): the function $f(s;u)$ is finite for any $u \in [0,\infty)$ and the asymptotic relation holds uniformly

$$f(s;u) \sim o\{u^{-p}\}, \qquad u \to \infty, \qquad p > 1. \tag{3.19}$$

Let us check the validity of this condition in the case of a GBM, which plays a central role due to Gibrat's rule. Recall that the pdf of a GBM is described by the log-normal distribution $\psi(x;\tau)$ given by (2.32), where τ is the counterpart of u, while $x = s/s_0$, where s_0 is the initial asset's value. One can easily show, that the function $\psi(x;\tau)$ is finite for any

$$x > 0 \quad \text{and} \quad x \neq 1 \qquad \Rightarrow \qquad s > 0 \quad \text{and} \quad s \neq s_0. \tag{3.20}$$

Besides, it is seen from (2.33–2.34), that $\psi(x;\tau)$ tends to zero exponentially, as $\tau \to \infty$

$$\psi(x;\tau) \sim o\left\{\exp\left(-\frac{\lambda^2}{4}\tau\right)\right\} \qquad \tau \to \infty$$

provided that $\lambda \neq 0$.

Thus, we have proved that in the case where firm's asset values obey a GBM, a steady-state density exists provided that condition (3.20) holds together with $\lambda \neq 0$. Noticing that, in the case $x = 1$ ($s = s_0$), the function $\psi(x;\tau)$ exhibits an integrable singularity in the vicinity of $\tau = 0_+$

$$\psi(x;\tau) \sim O\left\{\frac{1}{\sqrt{\tau}}\right\}, \qquad \tau \to 0_+, \qquad x = 1, \tag{3.21}$$

we conclude with

Proposition 3.3.1. *Under Assumptions 1–3, in the case where the firm's asset value is governed by a GBM, i.e., continuous Gibrat's law holds, and provided that the intensity of firms births is constant* ν*, it exists a steady-state mean density of firm's size for any* $s > 0$ *if and only if* $\lambda \neq 0$.

Later on, we will discuss in details the meaning of the condition $\lambda \neq 0$, from the point of view of the behavior of the realizations of the stochastic process $S(t)$ (2.10). Here we only notice that, if $\lambda > 0$, one can prove that the realizations of the stochastic process $S(t)$ tend, almost surely, to $+\infty$, as $t \to \infty$. The convergence of the improper integral (3.18) for any finite $s > 0$ follows. Analogously, if $\lambda < 0$, the realizations of $S(t)$ tend, almost surely, to zero, as $t \to \infty$, so that integral (3.18) converges as well.

3.4 Steady-State Density of Firm's Asset Values Obeying Gibrat's Law

Let us calculate the steady-state density (3.18), supposing that the initial asset value of all firms is the same and is equal to s_0. Due to the scale invariance of Gibrat's law, the pdf of a given firm's current asset value can be expressed in terms of the log-normal distribution (2.32) of the GBM (2.11) in the following way

$$f(s;s_0,t) = \frac{1}{s_0}\psi\left(\frac{s}{s_0};\frac{b^2t}{2}\right). \tag{3.22}$$

Substituting this in (3.18) and using the relation[3]

$$\begin{aligned} &\int_0^\infty \frac{1}{2\sqrt{\pi\tau}}\exp\left(-\frac{(v-c\tau)^2}{4\tau}-\mu\tau\right)d\tau \\ &= \frac{1}{\sqrt{c^2+4\mu}}\exp\left[\frac{1}{2}\left(cv-\sqrt{c^2+4\mu}\,|v|\right)\right], \qquad c^2+4\mu > 0, \end{aligned} \tag{3.23}$$

we obtain the following inverse relations for stochastically decaying processes ($\lambda < 0$) and for stochastically growing ones ($\lambda > 0$).

[3] This equality can be found in Bateman and Erdelyi (1954) together with the condition of validity $c^2 + 4\mu > 0$. We discuss here the particular case $\mu = 0$ for which the improper integral (3.23) uniformly converges provided that $c \neq 0$ or, equivalently, for $\lambda \neq 0$. We have derived the necessity of this condition for the convergence of the corresponding improper integral in the previous subsection.

Proposition 3.4.1. *Under the assumptions of Proposition 3.3.1, the steady-state mean density of firm's size exists and is such that*

$$g(s) = \frac{\nu}{|c|} \begin{cases} s^{-1}, & 0 < s < s_0, \\ s_0^{-\lambda} s^{\lambda-1}, & s > s_0, \end{cases} \qquad \lambda < 0, \tag{3.24}$$

which includes the balanced growth case $\lambda = -1$, while

$$g(s) = \frac{\nu}{c} \begin{cases} s_0^{-\lambda} s^{\lambda-1}, & 0 < s < s_0, \\ s^{-1}, & s > s_0, \end{cases} \qquad \lambda > 0. \tag{3.25}$$

It can be seen from the above relations that the steady-state mean density possesses two, lower and upper, power tails, which exchange their role as the parameter λ crosses the critical value $\lambda = 0$. Below, we will give a detailed explanation of this exchange, and in addition explain the values of the exponents of the upper and lower tails.

Here we note only that the upper tail of the mean density (3.24), corresponding to the stochastically decaying GBM ($\lambda < 0$) is

$$g(s) = \frac{\nu}{|c|} s_0^{-\lambda} s^{\lambda-1}, \qquad s > s_0, \quad \lambda < 0. \tag{3.26}$$

In the balanced case $\lambda = -1$, this corresponds to Zipf's law

$$g(s) = \frac{\nu}{|c|} s_0 \, s^{-2}, \qquad s > s_0. \tag{3.27}$$

Thus, we have derived Zipf's law from Gibrat's law and the balance condition, without requiring Gibrat's law to be broken in the vicinity of some small level $s_{\min} > 0$. Instead, we have taken into consideration the existence of births of firms at random instants $\{t_i\}$. Given the empirical evidence presented in Sect. 3.1, this approach seems much more natural than those based on Gabaix's supposition.

The existence of a possible randomness of the initial asset values s_0 does not change our main results and does not destroy Zipf's law. If, as a particular example, the pdf $f_0(s)$ of the random value s_0 of initial assets has a finite support, so that it is identically equal to zero for $s_0 > s_m$, where $s_m < \infty$ is some maximal possible initial asset value, then for any $s > s_m$, the exact Zipf's law holds:

$$g(s) = \frac{\nu}{|c|} E\left[s_0\right] \, s^{-2}, \qquad s > s_m. \tag{3.28}$$

Here, $E\left[s_0\right]$ denotes the expectation taken over the distribution of possible initial firm sizes. We will consider the case of random initial firm's sizes in more detail at the end of Chap. 5.

3.5 Mean Density of Firms Younger than Age t

For a better comprehensive understanding of power laws, and in particular of Zipf's law, which follows from Gibrat's law for the stochastic behavior of firm's asset values, let us consider the case of a finite time t. In this situation, the expression for the mean density of firm's asset values (3.16) reduces to

$$g(s,t) = \nu \int_0^t f(s;u)du. \tag{3.29}$$

It describes, for instance, the mean density of firm's asset values in some stable market ($\nu = \text{const.}$), if the statistical treatment takes into account only firms younger than some age t. Plots demonstrating the convergence of the mean density $g(s,t)$ given by (3.29) to the steady-state mean density $g(s)$ given by (3.24) with the growth of t when $\lambda = -1$ are depicted in Fig. 3.2.

Taking the annual volatility equal to $b = 0.2$, we find that $\tau = b^2t/2 = 10$ corresponds to a calendar time of $t = 500$ years, while $\tau = 100$, for which Zipf's

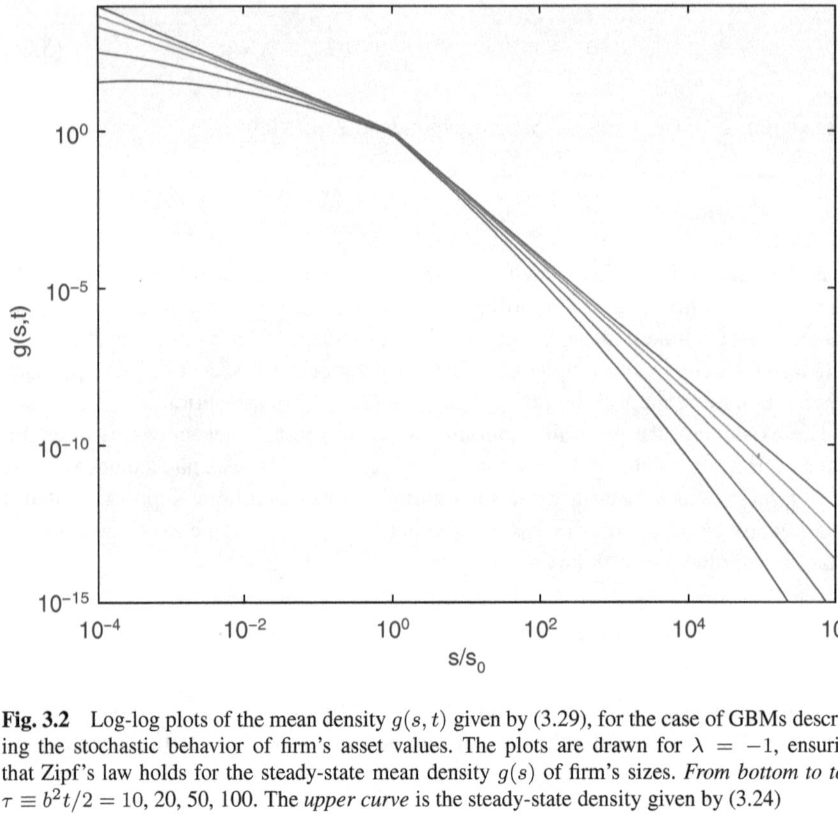

Fig. 3.2 Log-log plots of the mean density $g(s,t)$ given by (3.29), for the case of GBMs describing the stochastic behavior of firm's asset values. The plots are drawn for $\lambda = -1$, ensuring that Zipf's law holds for the steady-state mean density $g(s)$ of firm's sizes. *From bottom to top,* $\tau \equiv b^2t/2 = 10$, 20, 50, 100. The *upper curve* is the steady-state density given by (3.24)

law is recovered, corresponds to 5,000 years. It would thus seem that the convergence to Zipf's law is extremely slow, perhaps too slow to describe the real world. However, Schwarzkopf and Farmer recently reported that the monthly volatility of mutual fund sizes is about 0.25 (see Table 1 in Schwarzkopf and Farmer, 2008), corresponding to an annual volatility $b \simeq 0.87$. Accordingly, the characteristic time $t_b = 2/b^2$ is $t_b \simeq 2.67$ years and thus $\tau = t/t_b = 10$ corresponds to $t = 27$ years, while $\tau = 100$ corresponds to 267 years. This rate of convergence to Zipf's law is compatible with the empirical data described in Schwarzkopf and Farmer (2008).

3.6 Heuristic derivation of the origin of the power law distribution of firm sizes given by Gibrat's rule

The GBM resulting from Gibrat's law was shown above to be described by the log-normal distribution (3.22) of firm's asset values, which itself was demonstrated to be accurately approximated at large times by the power law

$$f(s; s_0, t) \sim s^{\lambda/2-1}. \tag{3.30}$$

How can we then understand intuitively that this power law transforms into another power law $g(s) \sim s^{\lambda-1}$ given by (3.26) for the steady-state mean density in the presence of firm's birth when $\lambda < 0$?

To better understand the transition from (3.30) to (3.26), let us study in more details the steady-state mean density $g(s)$ given by (3.18) in which $f(s;t) = f(s; s_0, t)$ is given by (3.22), itself derived from the log-normal distribution (2.32). For all $s > s_0$, we split the integral (3.18) into two parts

$$g(s) = \nu \left(\int_0^{4\tau_p\left(\frac{s}{s_0}\right)} f(s; s_0, u)du + \int_{4\tau_p\left(\frac{s}{s_0}\right)}^{\infty} f(s; s_0, u)du \right), \tag{3.31}$$

where $\tau_p(x)$ has been defined by (2.36) in such a way that for all $\tau \gtrsim \tau_p$, the asymptotics

$$f(s; s_0, \tau) \simeq \psi_0(\tau) s^{-1} \left(\frac{s}{s_0}\right)^{\lambda/2}, \tag{3.32}$$

holds uniformly in τ. Note that (3.32) refines (3.30) by providing the pre-factors $\psi_0(\tau)$ [given by expression (2.34)] of the asymptotic power law dependence in s.

Considering the first integral in the r.h.s. of equality (3.31), we get

$$\int_0^{4\tau_p\left(\frac{s}{s_0}\right)} f(s; s_0, u)du \leq 4\tau_p\left(\frac{s}{s_0}\right) \cdot \max_{u\in[0,4\tau_p]} f(s; s_0, u), \tag{3.33}$$

$$\sim \sqrt{\frac{\ln \frac{s}{s_0}}{\pi|\lambda|}} \, (s/s_0)^{1/2(|\lambda|-\lambda)-1}, \quad \text{as } s \gg s_0. \tag{3.34}$$

Using the asymptotic relation (3.32), the second integral in the r.h.s. of (3.31) reads

$$\int_{4\tau_p}^{\infty} f(s; s_0, u) du \sim s^{\lambda/2-1} \int_{4\tau_p}^{\infty} \psi_0(\tau) d\tau. \tag{3.35}$$

The integral in the r.h.s. of the last relation becomes

$$\int_{4\tau_p}^{\infty} \psi_0(\tau) d\tau = \frac{1}{|\lambda|} \operatorname{erfc}\left(|\lambda|\sqrt{\tau_p}\right), \tag{3.36}$$

$$\sim \exp\left(-\lambda^2 \tau_p\right), \qquad \tau_p \gg 1, \tag{3.37}$$

$$\sim \left(\frac{s}{s_0}\right)^{-|\lambda|/2}, \qquad s \gg s_0, \tag{3.38}$$

and

$$\int_{4\tau_p}^{\infty} f(s; s_0, u) du \sim s^{-(|\lambda|-\lambda)/2-1}, \qquad s \gg s_0. \tag{3.39}$$

As a consequence, for $\lambda < 0$, the first integral in the r.h.s. of (3.31) can be neglected for large s so that $g(s) \sim s^{\lambda-1}$.

Thus, roughly speaking, the transition from the asymptotic power law (3.30), for the log-normal distribution of any firm's asset value, to the power law (3.26) for the steady-state mean density of firm sizes, is due to the increase in time of the lower threshold above which the power law asymptotics (3.30) holds, which grows as $s_p \sim e^{\tau}$.

Appendix

Examples of Firm Birth Flows

The derivation of (3.14) and (3.16) holds for arbitrary stochastic births flows $\{t_i\}$ provided that the following assumptions hold: (1) the stochastic processes $S(t, t_i)$ and the births flow $\{t_i\}$ are independent, and (2) there exists a continuous mean birth rate $\nu(t)$. In this appendix, we present three examples of typical birth flow processes $\{t_i\}_{i\in\mathbb{N}}$ to provide a better understanding of the meaning of the mean rate $\nu(t)$.

Inhomogeneous Poisson Birth Flow

As a first example, we consider the inhomogeneous Poisson process $\{t_i\}_{i\in\mathbb{N}}$, assuming for simplicity, but without loss of generality, that the origin time of the economy and the time origin coincide: $t_0 = 0$. Let us divide the interval $u \in [0, t]$ into k adjacent intervals

$$u \in [u_k, u_{k+1}], \quad u_k := \frac{k}{n} t, \quad \Delta := u_{k+1} - u_k = \frac{t}{n}, \quad k = 0, \ldots n-1, \tag{3.40}$$

and assume that the number $\tilde{N}_k$ of births within each interval obeys the standard Poisson law, such that the expected number of births is equal to

$$\mathrm{E}\left[\tilde{N}_k\right] = \nu_k \Delta. \tag{3.41}$$

The characteristic function $\varphi_k(v)$ of the random number $\tilde{N}_k$ of births in the k-th interval is equal to

$$\varphi_k(v) := \mathrm{E}\left[e^{iv\tilde{N}_k}\right] = \exp\left[\nu_k \Delta (e^{iu} - 1)\right]. \tag{3.42}$$

By definition of Poisson processes, the numbers of births within different intervals (3.40) are independent random variables. As a consequence, the total number of births within the interval $(0, t)$, defined by

$$\tilde{N}(t) := \sum_{k=0}^{n-1} \tilde{N}_k, \tag{3.43}$$

admits the following characteristic function

$$\varphi(v; t) := \mathrm{E}\left[e^{iv\tilde{N}(t)}\right] = \prod_{k=0}^{n-1} \varphi_k(v) = \exp\left[(e^{iv} - 1) \sum_{k=0}^{n-1} \nu_k \Delta\right]. \tag{3.44}$$

Let $\nu(u)$ be some integrable function such that

$$\nu(t_k) := \nu_k. \tag{3.45}$$

In the limit $n \to \infty$, the sum in (3.44) converges to the integral

$$\lim_{n\to\infty} \sum_{k=0}^{n-1} \nu_k \Delta = \int_0^t \nu(u) du, \tag{3.46}$$

so that, at the limit, the characteristic function (3.44) reads

$$\varphi(v; t) = \exp\left[(e^{iv} - 1) \int_0^t \nu(u) du\right]. \tag{3.47}$$

Correspondingly, the expected value of the number of births within the interval $(0, t)$ is equal to

$$N(t) = -i \frac{\partial \varphi(v; t)}{\partial v}\bigg|_{v=0} = \int_0^t \nu(u) du. \tag{3.48}$$

In this process, the mean birth rate $\nu(t)$ of firm's birth is therefore nothing but the intensity of the inhomogeneous Poisson process.

Stationary Renewal Process

The firms birth flow

$$\{t_i\}, \qquad i = \ldots -2, -1, 0, 1, 2 \ldots \tag{3.49}$$

is a stationary Renewal Process (Karlin and Taylor, 1975, for instance), if the intervals

$$\tau_k := t_{k+1} - t_k, \qquad k = \ldots -2, -1, 0, 1, 2, \ldots \tag{3.50}$$

between two subsequent births are (positive) independent and identically distributed random variables. Let us denote by $\kappa(t)$ the pdf of each τ_k. We assume that the expected value of the random variable τ_k is finite and is equal to the integral

$$\mathrm{E}\left[\tau_k\right] = \int_0^\infty t\kappa(t)dt < \infty.$$

We define ν $(0 < \nu < \infty)$ as follows

$$\nu\,\mathrm{E}\left[\tau_k\right] = 1 \qquad \Rightarrow \qquad \nu = 1/\mathrm{E}\left[\tau_k\right]. \tag{3.51}$$

Let us consider, additionally, the random interval

$$\tau_+ = t_1, \tag{3.52}$$

where t_0 is the time interval from the origin of time $t = 0$ till the birthdate t_1 of the first firm (see Fig. 3.3).

In general, its pdf $\kappa_+(t)$ does not coincide with the pdf $\kappa(t)$ of the other intervals (3.50). In order to obtain the relation between the pdf's $\kappa_+(t)$ and $\kappa(t)$, we note that, due to the stationarity of the point process $\ldots t_{k-1},\ t_k,\ t_{k+1} \ldots$, the conditional

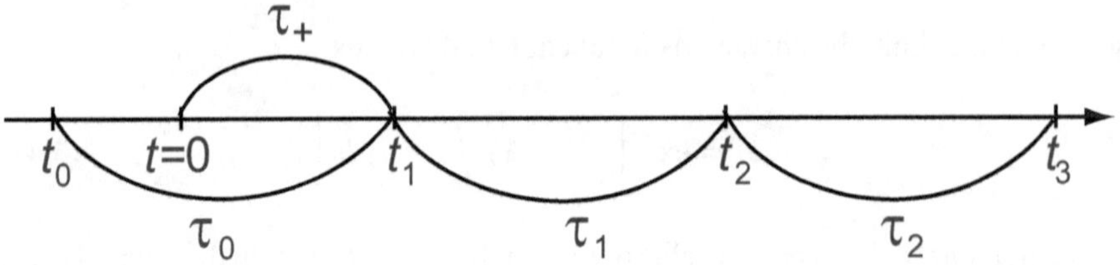

Fig. 3.3 Birthdates of firms and definition of τ_+ as the duration of the time interval beginning from the origin of time $t = 0$ till the birthdate t_1 of the first firm

pdf $\kappa_+(t|s)$ of the random durations of the intervals τ_+, under the condition that the duration of the corresponding whole (virtual) interval τ which includes $t = 0$ would be equal to a given s,[4] is the uniform distribution

$$\kappa_+(t|s) = \begin{cases} \frac{1}{s}, & 0 < t < s, \\ 0, & t < 0,\ t > s. \end{cases} \tag{3.53}$$

Averaging this uniform distribution over the statistics $\hat{\kappa}(t)$ of durations of the whole intervals $\tau(t_0)$ yields

$$\kappa_+(t) = \int_0^\infty \kappa_+(t|s)\hat{\kappa}(s)ds = \int_t^\infty \hat{\kappa}(s)\frac{ds}{s}. \tag{3.54}$$

Here, $\hat{\kappa}(t)$ is the pdf of the random durations of the interval $\tau(t_0)$, which is related to the pdf $\kappa(t)$ by

$$\hat{\kappa}(t) = \frac{t}{\mathrm{E}[\tau_k]}\kappa(t). \tag{3.55}$$

The factor t in the r.h.s. of (3.55) expresses the fact that the probability for the origin of time $t = 0$ to fall within the interval τ_0 of duration t is proportional to the duration of this interval times its pdf $\kappa(t)$. The constant $1/\mathrm{E}[\tau_k]$ ensures the normalization of $\hat{\kappa}(t)$. Putting (3.55) in (3.54) finally yields (Feller, 1966, for instance)

$$\kappa_+(t) = \frac{1}{\mathrm{E}\left[\tau_k\right]}\int_t^\infty \kappa(x)dx. \tag{3.56}$$

In view of relation (3.51), it reads

$$\kappa_+(t) = \nu\int_t^\infty \kappa(x)dx. \tag{3.57}$$

We will expand on the meaning of this pdf below. Here, we just mention that the expected value of the random variable τ_+ (3.52) exists only if

$$\mathrm{E}\left[\tau_k^2\right] = \int_0^\infty x^2\kappa(x)dx < \infty, \tag{3.58}$$

and is equal to

$$\mathrm{E}\left[\tau_+\right] = \frac{\nu}{2}\,\mathrm{E}\left[\tau_k^2\right]. \tag{3.59}$$

We will assume, in what follows, that inequality (3.58) holds.

In order to study the statistical properties of the Renewal Process $\{t_i\}_{i\in\mathbb{N}}$, two approaches are possible. The first one consists in studying the time duration until

---

[4] Where τ is defined as the waiting time $t_1 - t_0$, obtained by extending the process backward by one event.

the n-th birth, which amounts to studying the random sum

$$T(n) := \tau_+ + \sum_{k=1}^{n-1} \tau_k. \tag{3.60}$$

The cumulative distribution function of $T(n)$ is $K_n(t)$ given by

$$K_n(t) = \int_0^t \kappa_+(x) \otimes \underbrace{\kappa(x) \otimes \cdots \otimes \kappa(x)}_{n-1} dx, \quad K_1(t) = \int_0^t \kappa_+(x) dx, \tag{3.61}$$

where $\otimes$ denotes the convolution product.

Another approach consists in describing the properties of the random number of births $\tilde{N}(t)$ within the interval $[0, t]$. Let $P(n; t)$ be the probability that n births occur within the interval $[0, t]$. The cumulative distribution function $K_n(t)$ (3.61) of the random sum $T(n)$ (3.60) is related to the probability $P(n; t)$ of the random numbers $\tilde{N}(t)$ by the relation (Karlin and Taylor, 1975, for instance)

$$P(n; t) = K_n(t) - K_{n+1}(t) \quad (n \geqslant 1), \qquad P(0; t) = 1 - K_1(t). \tag{3.62}$$

In view of relation (3.9), we need the expected value $N(t)$ (3.10) of the number of births within the interval $(0, t)$. It follows from (3.62) that it is equal to

$$N(t) = \sum_{n=1}^{\infty} n\, P(n; t) = \sum_{n=1}^{\infty} K_n(t) \qquad (t > 0). \tag{3.63}$$

It is convenient to explore the properties of this sum by use of the Laplace transforms of its summands. The Laplace transform of some integrable function $Q(t)$ is denoted as

$$\hat{Q}(\rho) := \int_0^{\infty} Q(t)\, e^{-\rho t} dt. \tag{3.64}$$

Applying the Laplace transform to both sides of equality (3.63) and assuming that the series (3.63) converges uniformly within the interval $[0, \infty)$, we obtain

$$\hat{N}(\rho) = \sum_{n=1}^{\infty} \hat{K}_n(\rho). \tag{3.65}$$

The series in (3.65) converges absolutely for any $\rho > 0$, as we now show. The Laplace transform of $K_n(t)$ is [by (3.61)]

$$\hat{K}_n(\rho) = \frac{1}{\rho}\, \hat{\kappa}_+(\rho) \cdot \hat{\kappa}^{n-1}(\rho). \tag{3.66}$$

Since $\hat{\kappa}(\rho) < 1$ for $\rho > 0$, the r.h.s. of equality (3.65) is a convergent geometric series, whose summation yields

$$\hat{N}(\rho) = \frac{1}{\rho} \frac{\hat{\kappa}_+(\rho)}{1 - \hat{\kappa}(\rho)}. \tag{3.67}$$

To better understand the sense of this last relation, we introduce an important characteristic of the births flow $\{t_i\}_{i\in\mathbb{N}}$. Let $T > 0$ be some instant, while $i(T)$ is the number indexing the first birth occurring immediately after the instant T. In other words, $i - 1$ firm births occurred within the time interval $[0, T]$. Consider the time interval

$$\tau_+(T) = t_{i(T)} - T. \tag{3.68}$$

It is well-known that, under the conditions imposed above on the Renewal Process $\{t_i\}_{i\in\mathbb{N}}$ (stationarity and existence of first and second moments of the pdf of waiting times between births), the pdf $\kappa_+(t; T)$ of the random variable (3.68) admits the following limit

$$\lim_{T\to\infty} \kappa_+(t; T) = \kappa_+(t), \tag{3.69}$$

where $\kappa_+(t)$ is given by equality (3.57). In words, the limit (3.69) means that, as $T \to \infty$, the Renewal Process becomes "truly stationary," in the sense that the statistical properties of the time intervals (3.50) do not depend on i, and in addition the statistical properties of the intervals (3.68) do not depend on the current time T. The essence of the demonstration of formula (3.57) is based on the formulation that views T falling equiprobably within any interval $[t_k, t_{k+1}]$. A Renewal Process $\{t_i\}_{i\in\mathbb{N}}$, for which the pdf of random interval (3.52) obeys relation (3.57), can be called a *stationary Renewal Process*.

Applying Laplace transform to the pdf (3.57) yields

$$\hat{\kappa}_+(\rho) = \frac{\nu}{\rho} \left[1 - \hat{\kappa}(\rho)\right]. \tag{3.70}$$

After substitution into equality (3.67), we finally get

$$\hat{N}(\rho) = \frac{\nu}{\rho^2} \qquad \Rightarrow \qquad N(t) = \nu t \qquad (t > 0). \tag{3.71}$$

This last relation means that, if the births flow $\{t_i\}_{i\in\mathbb{N}}$ is a stationary Renewal Process, then, irrespective of the pdf $\kappa(t)$ of the random intervals (3.50), the mean birth rate is constant, and is defined by equality (3.51). Then, the mean density (3.16) of firms assets values reads

$$g(s, t) = \nu \int_0^t f(s; u)du. \tag{3.72}$$

This relation remains true even for a pure periodic birth flow, for which the pdf of intervals (3.50) is singular

$$\kappa(t) = \nu\delta(\nu t - 1), \tag{3.73}$$

where $\delta(x)$ is Dirac's delta function. As seen from (3.57), such kind of periodic birth flow with period $1/\nu$ becomes stationary if the pdf of the initial interval τ_+ (3.52) is uniform:

$$\kappa_+(t) = \nu\,\mathbf{1}(1 - \nu t). \tag{3.74}$$

Rarefied Poisson Flow

We here illustrate the properties of Renewal Processes for the particular case of rarefied Poisson processes. We first need to recall some properties of homogeneous Poisson point processes, for which the pdf of interval durations (3.50) are exponential

$$\kappa(t;\nu) = \nu e^{-\nu t}, \qquad t > 0. \tag{3.75}$$

Due to the Markovian property of Poisson point processes, it is easy to show that the pdf (3.57) of the initial interval (3.52) coincides with the exponential distribution (3.75):

$$\kappa_+(t;\nu) = \kappa(t;\nu) = \nu e^{-\nu t}. \tag{3.76}$$

Correspondingly, one may replace the sum (3.60) by

$$T(n;\nu) = \sum_{k=0}^{n-1} \tau_k, \tag{3.77}$$

where the summands are i.i.d and possess some pdf $\kappa(t;\nu)$. As is well-known from the theory of Poisson point processes, the pdf of the sum (3.77) is equal to

$$\kappa_n(t;\nu) = \nu \frac{(\nu t)^{n-1}}{\Gamma(n)} e^{-\nu t}, \qquad t > 0. \tag{3.78}$$

Thus, the cumulative distribution function of the random sum (3.77) is

$$K_n(t;\nu) = \int_0^t \kappa_n(x;\nu)dx. \tag{3.79}$$

Accordingly, the mean number of births of the Poisson point process, within the time interval $[0,t]$, is equal to

$$N(t) = \sum_{n=1}^{\infty} \int_0^t \kappa_n(x;\nu)dx.$$

Exchanging the order of the summation and integration, and taking into account the identity

$$\sum_{n=1}^{\infty} k_n(x;\nu) = \nu e^{-\nu t} \sum_{n=1}^{\infty} \frac{(\nu t)^{n-1}}{(n-1)!} = \nu e^{-\nu t} \cdot e^{\nu t} = \nu, \tag{3.80}$$

we finally obtain

$$N(t) = \nu t. \tag{3.81}$$

We are now armed to study rarefied Poisson processes. We will call the point process $\{t_i\}_{i\in\mathbb{N}}$ a rarefied Poisson point process of order m, if the waiting time $\tau_k(m)$ between two successive births is given by

$$\tau_k(m) = \sum_{j=1}^{m} \tau_{(k-1)m+j}, \tag{3.82}$$

where $\tau_{(k-1)m+j}$ denotes the duration of the $(k-1)m+j$ time interval of a Poisson process with intensity ν. By construction, the pdf's of these sums are equal to $\kappa_m(t;\nu)$ given by (3.78). By substitution into (3.56), and taking into account that the expected value of the random sum (3.82) is equal to

$$\mathrm{E}\left[\tau_k(m)\right] = \frac{m}{\nu}, \tag{3.83}$$

we obtain the pdf of the initial interval $\tau_+(m)$ of a stationary rarefied Poisson process

$$\kappa_+(t;\nu|m) = \frac{1}{m}\sum_{j=1}^{m} \kappa_j(t;\nu). \tag{3.84}$$

The mean rate of births of the rarefied Poisson process can be calculated as follows. Analogously to (3.63), the expected number of births within the time interval $[0,t]$ is equal to

$$N(t;m) = \sum_{n=0}^{\infty} K_n(t;m), \qquad t > 0, \tag{3.85}$$

where

$$\begin{aligned} K_n(t;m) &= \int_0^t \kappa_+(x;\nu|m) \otimes \kappa_{mn}(x;\nu)dx, \quad n > 0, \\ K_0(t;m) &= \int_0^t \kappa_+(x;\nu|m)dx. \end{aligned} \tag{3.86}$$

By substitution of expression (3.84) and by use of the semigroup property of the Poisson pdf's (3.78) which reads

$$\kappa_m(t;\nu) \otimes \kappa_n(t;\nu) = \kappa_{m+n}(t;\nu) \qquad \forall m,n = 1,2,\ldots, \tag{3.87}$$

(3.85) can be rewritten in the form

$$N(t;m) = \frac{1}{m}\int_0^t \sum_{n=0}^{\infty}\sum_{j=1}^{m} \kappa_{nm+j}(x;\nu)dx. \tag{3.88}$$

Obviously,

$$\sum_{n=0}^{\infty}\sum_{j=1}^{m}\kappa_{nm+j}(x;\nu)=\sum_{n=1}^{\infty}\kappa_n(x;\nu)=\nu,$$

so we have

$$N(t)=\frac{\nu}{m}\,t. \tag{3.89}$$

This last relation quantifies how the rarefaction of a Poisson process with mean rate ν creates a stationary rarefied Poisson process of order m, whose mean rate is m times smaller. In order to keep the mean rate constant, while changing the order m of the rarefied Poisson process, we have to replace the mean rate ν of the original Poisson process by $\nu \cdot m$. Figure 3.4 shows the rescaled pdf's (3.84) of the initial interval $\tau_+(m)$ for different values of m. While the stationary rarefied Poisson process becomes more and more similar to a periodic point process when m increases, the pdf (3.84) converges, as $m \to \infty$, to the uniform distribution (3.74).

The situation drastically changes, at least for $t \lesssim 1/\nu$, if the distribution of the initial interval $\tau_+(m)$ (3.50) is different from (3.84). To illustrate this remarkable fact, let us replace $\kappa_+(t;\nu|m)$ by the pdf $\kappa_m(t;\nu)$ (3.78) of any subsequent interval $\tau_k(m)$ of the rarefied Poisson process. To make the exposition slightly more gen-

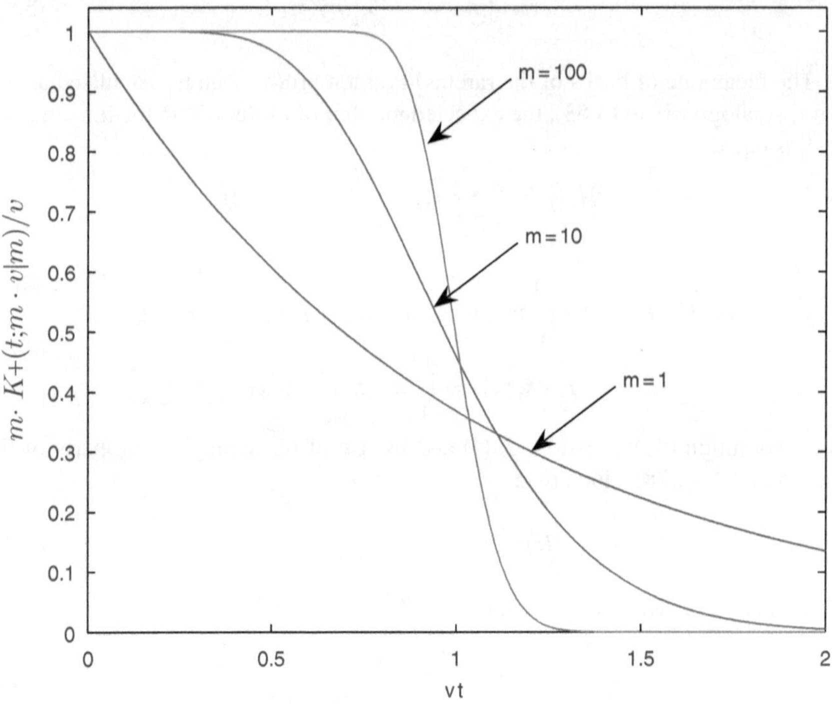

Fig. 3.4 Plots of rescaled pdf's $m\kappa_+(t;m\nu|m)$ of the duration of the initial interval $\tau_+(m)$, for rarefied Poisson point processes of orders $m = 1$, 10, 100

eral, we interpret $\kappa_m(t;\nu)$ (3.78) as a Gamma distribution, so that the parameter m can take arbitrary positive real values. Recall that $m = 1$ corresponds to a Poisson process, which describes mathematically a pure stochastic dissemination of the points $\{t_i\}_{i\in\mathbb{N}}$. When $m > 1$, there is some repulsion between successive points, while points aggregate into clusters for $m \in (0,1)$.

In this case, one has to substitute into (3.85)

$$K_n(t;m) = \int_0^t \kappa_{(n+1)m}(x;\nu)dx. \tag{3.90}$$

Accordingly, the expected number $N(t;\nu)$ of points within the interval $(0,t)$ is equal to

$$N(t;\nu|m) = \int_0^t \nu(x;\nu|m)dx, \tag{3.91}$$

where the mean rate

$$\nu(t;\nu|m) = \sum_{n=1}^{\infty} \kappa_{nm}(t;\nu) \tag{3.92}$$

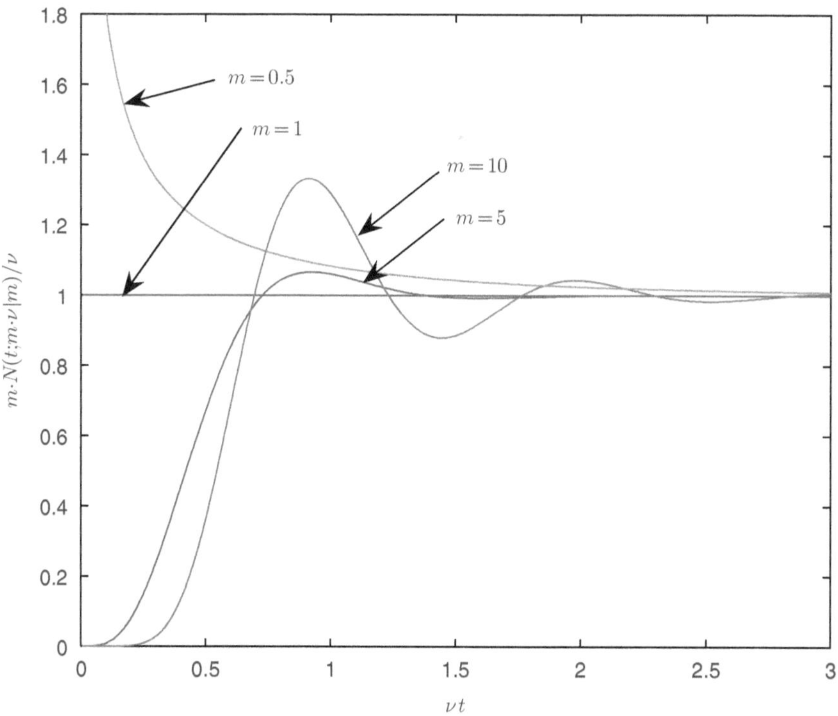

Fig. 3.5 Graphs of the rescaled mean rates (3.92) for $m = 0.5, 1, 5, 10$

is not constant. For instance

$$
\begin{aligned}
N(t;\nu|1/2) &= \nu\left[1+\frac{e^{-\nu t}}{\sqrt{\pi\nu t}}+\mathrm{erf}(\sqrt{\nu t})\right], \\
N(t;\nu|1) = \nu, &\qquad N(t;\nu|2) = \nu e^{-\nu t}\sinh(\nu t), \\
N(t;\nu|4) &= \frac{\nu}{2}e^{-\nu t}\left[\sinh(\nu t)-\sin(\nu t)\right].
\end{aligned}
\tag{3.93}
$$

Figure 3.5 depicts the mean rate (3.92) of non-stationary rarified Poisson processes, for different orders m. All the mean rates tend, for large values of νt, to a constant limit. At the same time, the repulsion between events associated with $m > 1$ makes the mean rate of birth smaller than the limiting value for not large νt. In contrast, clusters appear for $0 < m < 1$ and give rise to a singularity of the mean rate in the vicinity of $t = 0$.

Chapter 4
Useful Properties of Realizations of the Geometric Brownian Motion

In the previous chapter, we introduced the mean density of firm's asset values (3.18), taking into account the flow of firm's births $\{t_i\}$. We provided a preliminary analysis of the properties of the mean density of firm sizes. In order to understand more deeply the roots of the power laws (3.24) and (3.25), and at the same time the basis of Zipf's law (3.27) and (3.28), we discuss in detail in this chapter the statistical properties of the realizations of the GBM (2.11).

4.1 Relationship Between the Distributions of Firm's Mean Ages and Sizes

Let us first describe the intimate connection existing between the steady-state mean density $g(s)$ given by (3.18) and the stochastic behavior of the asset value $S(t)$ of some given firm.

We have defined $f(s;t)$ as the pdf of the current asset value $S(t)$ of some firm which was born at the initial instant $t_0 = 0$. The corresponding complementary cdf (or survival distribution) $\bar{F}(s;t)$ is equal to

$$\bar{F}(s;t) = \mathrm{E}[\mathbf{1}(S(t) - s)], \tag{4.1}$$

where $\mathbf{1}(z)$ is the unit step function, equal to one for $z \geq 0$ and zero for $z < 0$.

Let us introduce the integral

$$\int_0^{\infty} \mathbf{1}(S(t) - s)dt = \sum_k T_k, \tag{4.2}$$

where the T_k's are the durations of time intervals for which $S(t) > s$. This integral or sum in (4.2) converges for $\lambda < 0$. Figure 4.1 shows a schematic illustration of equality (4.2). The T_k's can be thought of as the waiting times between successive crossings of the level s. Averaging both sides of equality (4.2) over the statistical ensemble of the diffusion process $S(t)$, we obtain

A. Saichev et al., *Theory of Zipf's Law and Beyond*,
Lecture Notes in Economics and Mathematical Systems 632,
DOI 10.1007/978-3-642-02946-2_4, © Springer-Verlag Berlin Heidelberg 2010

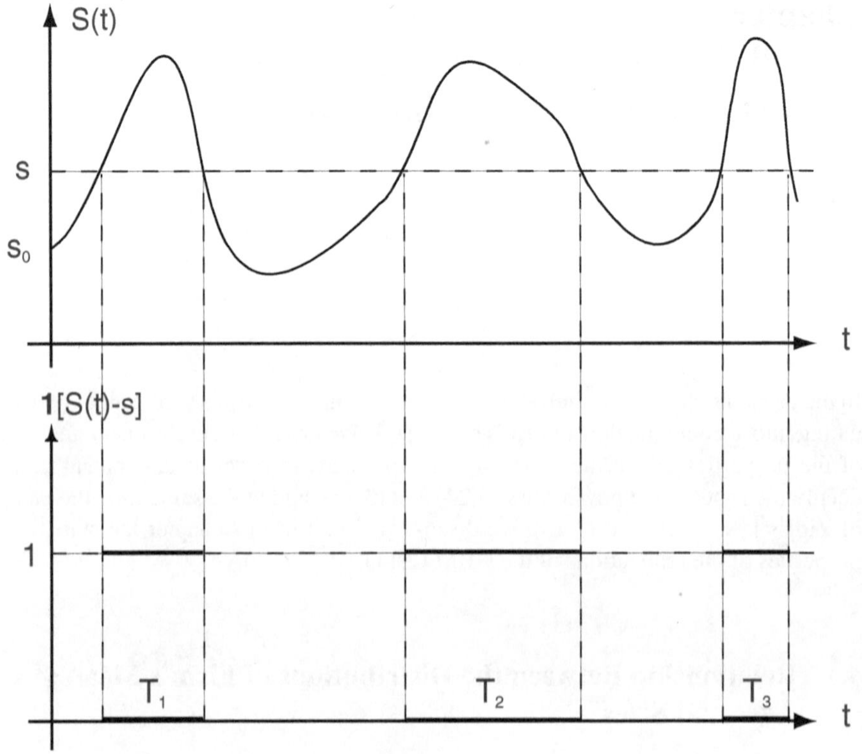

Fig. 4.1 Schematic illustration of equality (4.2)

$$\int_0^\infty \bar{F}(s;t)dt = T(s), \tag{4.3}$$

where

$$T(s) := \mathrm{E}\left[\sum_k T_k\right] = \sum_k \mathrm{E}\left[T_k\right] \tag{4.4}$$

is the mean of the total duration of all time intervals for which $S(t) > s$. In what follows, it will be convenient to use the probability density distribution of mean time interval durations

$$\theta(s) = -\frac{dT(s)}{ds} = \int_0^\infty f(s;t)dt. \tag{4.5}$$

Note that $\theta(s)$ differs only by the factor ν from the steady-state mean density $g(s)$ given by (3.18). When the intensity of firm's birth ν is constant, this provides a new interpretation of Zipf's law (3.27). Rewritten in terms of the mean total duration $T(s)$ given by (4.4) for $\lambda = -1$, it takes the form

$$T(s) = \frac{s_0}{|c|s}, \qquad s > s_0. \tag{4.6}$$

In other words, if the stochastic behavior of a firm's asset value obeys the balanced Gibrat's law, then the overall mean duration $T(s)$ ($s > s_0$) of time intervals when the firm has an asset value $S(t)$ exceeding a given level s, satisfies Zipf's law. It is worthwhile pointing out that Zipf's law does not mean that the larger the asset value s of some firm, the smaller the time its asset value remains larger than s. We will see in what follows that it means only that the larger the value of s, the smaller is the fraction of firms having an asset value larger than s.

4.2 Mean Growth vs. Stochastic Decay

As follows from (2.29), the mean value of the normalized GBM is equal to

$$\mathrm{E}\left[X(t,c,b)\right] = e^{(1+\lambda)\tau}. \tag{4.7}$$

and is growing (for $\lambda > -1$) even if $X(t,c,b)$ is stochastically decreasing, as occurs for $c < 0 \iff \lambda < 0$. It thus seems that, for $\lambda \in [-1, 0)$, there is a discrepancy between our terminology and the behavior of the mean value of the GBM.

To elucidate this paradox, consider the continuous limit of an hypothetical capital taxation system imposed on firms, which works as follows. At equidistant instants $t_m = m\Delta$, a firm has to pay a kind of capital tax equal to the fraction κ of its asset value S at that instant. Assuming a discount rate of zero (or equivalently subtracting the discount rate from the drift c of the GBM describing the time evolution of the firm value), the total tax revenue from an original instant $t_0 = 0$ at which the firm was established over the total life of the firm is equal to

$$\mathcal{T} = \kappa \sum_{m=1}^{\infty} S(m\Delta). \tag{4.8}$$

Bankruptcies could be taken into account in this formally infinite sum by simply writing that $S(m\Delta) = 0$ beyond the firm's death.

Let us assume that Δ is smaller than the characteristic time over which the firm's asset value $S(t)$ varies significantly, while $\kappa \ll 1$ so that the influence of taxation on the behavior of the firm's asset value $S(t)$ is negligible. One may then replace the above sum (4.8) by the integral

$$\mathcal{T} \simeq \frac{\kappa}{\Delta} \int_0^{\infty} S(t)dt. \tag{4.9}$$

Let us explore the statistical properties of the integral (4.9), replacing it, for convenience of analysis, by the integral over the dimensionless time τ defined by (2.18) of the GBM defined by (2.17), that we denote by

$$A(\lambda) := \int_0^\infty \chi(\tau,\lambda)d\tau, \tag{4.10}$$

where λ is defined by (2.19). The process $\chi(\tau,\lambda)$ satisfies the stochastic equation

$$d\chi(\tau,\lambda) = (1+\lambda)\chi(\tau,\lambda)d\tau + \chi(\tau,\lambda)d\omega(\tau). \tag{4.11}$$

Consider the auxiliary stochastic equation

$$dB(\tau,\lambda) = [1+(1+\lambda)B(\tau,\lambda)]d\tau + B(\tau,\lambda)d\omega(\tau), \quad B(\tau=0,\lambda)=0. \tag{4.12}$$

Let represent its solution in the form

$$B(\tau,\lambda) = \chi(\tau,\lambda)C(\tau,\lambda). \tag{4.13}$$

Substituting it into the previous equation and using the stochastic equation (4.11) yields the equation for $C(\tau,\lambda)$ as

$$dC(\tau,\lambda) = \chi^{-1}(\tau,\lambda)d\tau, \quad C(\tau=0,\lambda)=0. \tag{4.14}$$

Its solution reads

$$C(\tau,\lambda) = \int_0^\tau \frac{d\tau'}{\chi(\tau',\lambda)}. \tag{4.15}$$

Using (4.13), we obtain

$$B(\tau,\lambda) = \int_0^\tau \frac{\chi(\tau,\lambda)}{\chi(\tau',\lambda)}d\tau'. \tag{4.16}$$

Because $\chi(\tau,\lambda)$ enjoys the following symmetric property

$$\frac{\chi(\tau,\lambda)}{\chi(\tau',\lambda)} \overset{law}{=} \chi(\tau-\tau',\lambda), \tag{4.17}$$

we have

$$\int_0^\tau \frac{\chi(\tau,\lambda)}{\chi(\tau',\lambda)}d\tau' \overset{law}{=} \int_0^\tau \chi(\tau-\tau',\lambda)d\tau' = \int_0^\tau \chi(\tau',\lambda)d\tau'. \tag{4.18}$$

This gives

$$B(\tau,\lambda) \overset{law}{=} \int_0^\tau \chi(\tau',\lambda)d\tau'. \tag{4.19}$$

It follows from the last relation that, if $\int_0^\tau \chi(\tau',\lambda)d\tau'$ converges in probability to a random value $A(\lambda)$,

$$\lim_{T\to\infty}\int_0^T \chi(\tau',\lambda)d\tau' = A(\lambda), \tag{4.20}$$

then the pdf $\varepsilon(r;\lambda)$ of the random value $A(\lambda)$, defined by (4.10), coincides with the following limit

$$\varepsilon(r;\lambda) := \lim_{\tau\to\infty}\varepsilon(r;\lambda,\tau), \tag{4.21}$$

where $\varepsilon(r;\lambda,\tau)$ is the pdf of the solution $B(\tau,\lambda)$ to the stochastic equation (4.12). A standard result of the theory of stochastic processes is that $\varepsilon(r;\lambda,\tau)$ satisfies the diffusion equation

$$\frac{\partial\varepsilon}{\partial\tau} + \frac{\partial\varepsilon}{\partial r} + (1+\lambda)\cdot\frac{\partial(r\varepsilon)}{\partial r} = \frac{\partial^2(r^2\varepsilon)}{\partial r^2}, \qquad \varepsilon(r;\lambda,0) = \delta(r). \tag{4.22}$$

Defining the stationary probability flux

$$\Omega(r;\lambda) = [1+(1+\lambda)r]\,\varepsilon(r,\lambda) - \frac{d[r^2\varepsilon(r;\lambda)]}{dr}, \tag{4.23}$$

then, the limiting pdf (4.21), if it exists, must be the solution to the stationary equation

$$\Omega(r,\lambda) = \Omega_\infty = \text{const.}, \tag{4.24}$$

where Ω_∞ is the value of the probability flux at infinity. It is easy to show that, if $\lambda < 0$, then (4.24) with (4.23) has a non-degenerate solution, corresponding to a zero flux $\Omega_\infty = 0$. This solution reads:

$$\varepsilon(r,\lambda) = \frac{1}{\Gamma(|\lambda|)}\left(\frac{1}{r}\right)^{1-\lambda}\exp\left(-\frac{1}{r}\right), \qquad \lambda < 0. \tag{4.25}$$

Recall that $\varepsilon(r,\lambda)$ is the pdf of the random value $A(\lambda)$ given by (4.10).

We can now resolve the paradox introduced at the beginning of this section and justify the use of the term "stochastically decaying" in reference to the GBM in the case $\lambda < 0$. In this goal, we note that $A(\lambda)$ has the transparent geometrical sense of being the random area under the GBM realization $\chi(\tau,\lambda)$ (see Fig. 4.2). The existence of the non-degenerate pdf (4.25) then means that the random area under the realization $\chi(\tau,\lambda<0)$ is almost surely finite which implies that $\chi(\tau,\lambda<0)$ decays faster than $1/\tau$ as $\tau\to\infty$.

4.3 Geometrically Transparent Definitions of Stochastically Decaying and Growing Processes

In the two previous chapters, we have introduced and referred to the notion of stochastically decaying (or growing) GBMs defined by the condition $c < 0$ ($c > 0$). In the previous section, we discussed the finiteness of the areas of the domains

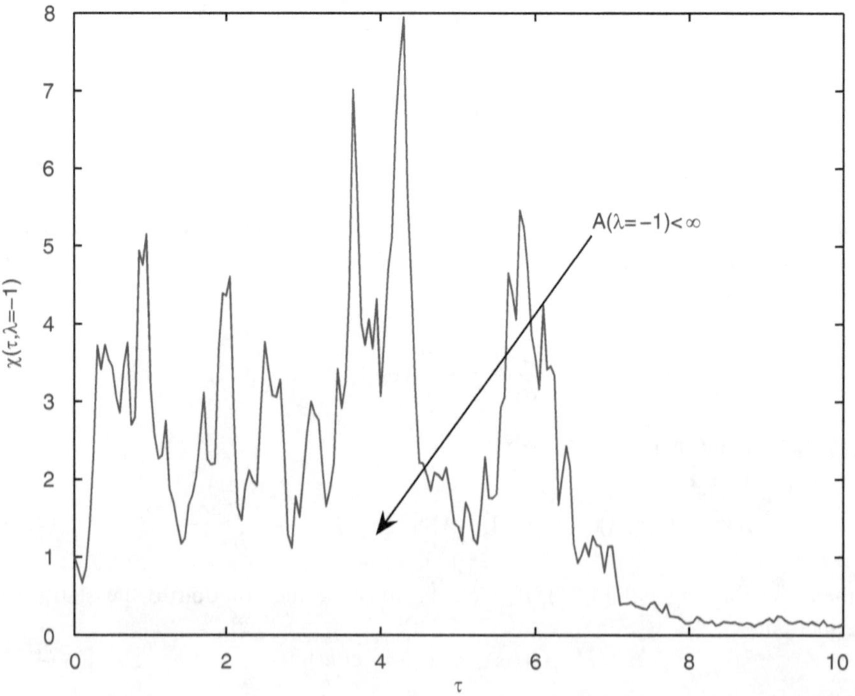

Fig. 4.2 A plot of one realization of the balanced GBM $\chi(\tau, \lambda = -1)$ defined in (2.17), illustrating the stochastic decay. Although the mean value of a balanced GBM is constant so that the mean area under its realization is equal to infinity, the random area under any realization of balanced GBMs is almost surely finite

delineated by the trajectories of the realizations of stochastically decaying GBMs. This suggests the following more geometrically transparent and precise definition of stochastically decaying and growing processes.

Definition 4.3.1. A non-negative diffusion process $\{S(t)\}_{t \in \mathbb{R}_+}$ is *stochastically decaying* if the area of the domain under its trajectory is almost surely finite:

$$\Pr\left\{\int_0^{\infty} S(t)dt < \infty\right\} = 1. \tag{4.26}$$

Consider as an example the case of the balanced GBM ($\lambda = -1$). As follows from expression (4.25), the pdf of the random areas under the realizations of balanced GBM samples $\chi(\tau, \lambda = -1)$ is equal to

$$\varepsilon(r, \lambda = -1) = \frac{1}{r^2} \exp\left(-\frac{1}{r}\right). \tag{4.27}$$

The mean of this distribution does not exist in compliance with the balance condition

$$\mathrm{E}[\chi(\tau,-1)] = 1 \quad \Rightarrow \quad \mathrm{E}[A(-1)] = \int_0^\infty \mathrm{E}[\chi(\tau,-1)]d\tau = \infty. \tag{4.28}$$

Nevertheless, as shown in the previous section, the area under $\chi(\tau, \lambda = -1)$ is almost surely finite, so that the balanced GBM is stochastically decaying.

We have already remarked that the pdf (4.27) expresses a kind of Zipf's law for the total tax revenue $\mathcal{T}$ defined by expression (4.8), when the behavior of the firm's asset value obeys the balanced Gibrat's law. Specifically, the complementary cdf (survival distribution) of the random areas under realizations $\chi(\tau,-1)$ of the balanced GBM is given by

$$\int_r^\infty \varepsilon(r,-1)dr = 1 - \exp\left(-\frac{1}{r}\right), \tag{4.29}$$

and decays as $\simeq 1/r$ for $r \gg 1$.

Relation (2.23) implies that the finiteness of the area under realizations of the GBM $\chi(\tau, \lambda < 0)$ is equivalent to the finiteness of the area under the inverse of the realizations of the diffusion process $\chi(\tau, \lambda > 0)$. This suggests a geometrically transparent notion of stochastically growing processes:

Definition 4.3.2. A positive diffusion process $\{S(t)\}_{t\in\mathbb{R}_+}$ is *stochastically growing* if

$$\Pr\left\{\int_0^\infty \frac{dt}{S(t)} < \infty\right\} = 1. \tag{4.30}$$

4.4 Majorant Curves of Stochastically Decaying Geometric Brownian Motion

In order to obtain a deeper understanding of Zipf's law (3.27) for the mean steady-state density $g(s)$ given by (3.18) of firm's asset values, we explore in more detail the statistical properties of the realizations of the GBM defined by (2.11).

Definition 4.4.1. Given a diffusion process $\{S(t)\}_{t\in\mathbb{R}_+}$, the mapping $M(t,p)\colon \mathbb{R}_+ \times (0,1) \mapsto \mathbb{R}$ is called a *majorant curve* at probability level p if

$$\Pr\left[S(t) < M(t,p); \forall t \in [0,\infty)\right] = p. \tag{4.31}$$

For the Wiener process, a particularly useful quantity is the probability that the process never touches the linearly growing boundary:

$$P(x,V) := \Pr\{bW(t) < \ln x + vt | t \in [0,\infty)\}, \qquad x > 1,\ v > 0. \tag{4.32}$$

We have introduced the dimensionless parameter

$$V := vt_b, \tag{4.33}$$

where t_b is defined in (2.18). It is well-known (see, for instance, Borodin and Salminen, 2002, p. 251) that

$$P(x, V) = 1 - x^{-V}. \tag{4.34}$$

In view of the statistical symmetry of the Wiener process, the following inverse relationship is true:

$$\Pr\{bW(t) > \ln x - vt | t \in [0, \infty)\} = P(x, -V), \qquad x < 1,\ v > 0. \tag{4.35}$$

With reference to the firm's current asset value (2.9), expression (4.32) with (4.34) means that there is a family of majorant curves

$$M(\tau, p|V) := s_0 \left(\frac{1}{1-p}\right)^{1/V} e^{(\lambda+V)\tau}, \qquad V > 0, \tag{4.36}$$

such that, with a given probability p, the firm's current asset value $S(t)$ will never exceed the majorant curve (4.36).

In the case of the stochastically decaying GBMs ($\lambda < 0$) under discussion, one may also choose $\lambda + V$ to be negative so that the majorant curve (4.36) is exponentially decaying. In particular, half of all realizations of the balanced ($\lambda = -1$) GBM defined by (2.11) are situated under the majorant curve

$$M(\tau, 1/2\,|1/2) = 4e^{-\tau/2}. \tag{4.37}$$

A plot of this exponentially decaying curve, and one example from half of the realizations of the GBM (2.11) that always remain beneath the given majorant curve, are depicted in Fig. 4.3.

4.5 Maximal Value of Stochastically Decaying Geometric Brownian Motion

Let us rewrite expression (4.36) in another form

$$M = s \cdot e^{(\lambda+V)\tau}, \tag{4.38}$$

where $s > s_0$ is the maximal value (if $\lambda < 0$, $V < |\lambda|$) of the majorant curve, which is related to the probability level p by the equality

$$s = s_0 \left(\frac{1}{1-p}\right)^{1/V}. \tag{4.39}$$

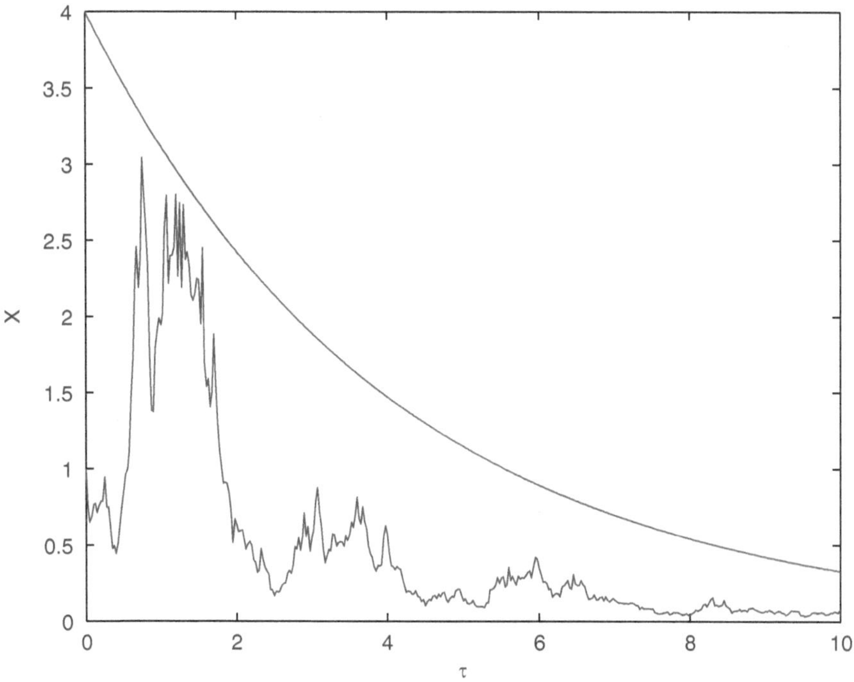

Fig. 4.3 Plots of the exponentially decaying majorant (4.37) and one example from half of the realizations of the balanced GBM (2.11) that are always beneath the majorant curve (4.37)

One may interpret this equality, for given s and V, as an equation in p. An obvious solution to this equation is

$$p = p_+(s, s_0, V) = 1 - \left(\frac{s_0}{s}\right)^V, \qquad s > s_0. \tag{4.40}$$

For the case $V = -\lambda$ ($\lambda < 0$), the r.h.s. of the last relation has a transparent geometrical interpretation. Indeed, by replacing $V = -\lambda$ in the r.h.s. of equality (4.38), we find that the majorant curve reduces to the constant $M = s$. This means in turn that

$$P_+(s, s_0, \lambda) := p_+(s, s_0, -\lambda) = 1 - \left(\frac{s_0}{s}\right)^{-\lambda}, \qquad s > s_0, \tag{4.41}$$

is the cdf of the maximal values

$$S_+ := \max_{t\in[0,\infty)} S(t) \tag{4.42}$$

over realizations of stochastically decaying asset value $S(t)$'s (for the case under discussion characterized by $\lambda < 0$). In other words,

$$P_{+}(s, s_0, \lambda) = \Pr\{S_{+} < s\}. \tag{4.43}$$

The corresponding complementary cdf is equal to

$$Q_{+}(s, s_0, \lambda) := 1 - P_{+}(s, s_0, \lambda) = \left(\frac{s_0}{s}\right)^{-\lambda}, \quad s > s_0, \quad \lambda < 0. \tag{4.44}$$

Taking $\lambda = -1$ (balanced Gibrat's law) in (4.44) leads to Zipf's law as shown in the previous chapter, together with the result that the complementary cdf of absolute maximal values of the realizations of the firm's asset value $S(t)$ obey

$$Q_{+}(s, s_0, -1) = \frac{s_0}{s}, \qquad s > s_0. \tag{4.45}$$

Note in conclusion that one may interpret $Q_{+}(s, s_0, \lambda)$ given by (4.44) as the probability that the realization $S(t)$ crosses a given level $s > s_0$. In the case $\lambda < 0$, the stochastically decaying realization of the firm's asset value $S(t)$ necessarily crosses any level below s_0, so we have $Q_{+}(s, s_0, \lambda) = 1$ for $s \in [0, s_0]$. Thus, the final expression for the probability $Q_{+}(s, s_0, \lambda)$ of crossing any level $s > 0$ by the stochastically decaying GBM $S(t)$ is

$$Q_{+}(s, s_0, \lambda) = \begin{cases} 1, & 0 < s < s_0, \\ \left(\frac{s_0}{s}\right)^{-\lambda}, & s > s_0, \end{cases} \qquad \lambda < 0. \tag{4.46}$$

4.6 Extremal Properties of Realizations of Stochastically Growing Geometric Brownian Motion

For stochastically decaying GBMs ($\lambda < 0$), and especially when the balanced Gibrat's law $\lambda = -1$ holds, the discussion of the previous chapter seems sufficient for substantiating Zipf's laws (3.27), (4.6), (4.29), (4.45). Nevertheless, to gain a comprehensive understanding of Zipf's and related power laws inherent to diffusion processes obeying Gibrat's law, we discuss in detail the opposite case of stochastically growing GBMs, corresponding to $\lambda > 0$.

We first introduce the notion of a *minorant curve*

Definition 4.6.1. Given a diffusion process $\{S(t)\}_{t \in \mathbb{R}_+}$, the mapping $m(t, p)\colon \mathbb{R}_+ \times (0, 1) \mapsto \mathbb{R}$ is called a *minorant curve* at probability level p if

$$\Pr\left[S(t) > m(t, p); \forall t \in [0, \infty)\right] = p. \tag{4.47}$$

Expressions (4.35) with (4.34) imply that one can define minorant curves

$$m(\tau|V) := s e^{(\lambda - V)\tau}, \qquad s < s_0, \tag{4.48}$$

such that, with probability

$$p_-(s, s_0, V) := 1 - \left(\frac{s}{s_0}\right)^V, \qquad s < s_0, \tag{4.49}$$

the current asset value $S(t)$ will never touch the minorant curve defined by (4.48). If we choose $0 < V < \lambda$, then the minorant curve (4.48) is exponentially growing. In particular, half of the realizations of the GBM (2.11), corresponding to $\lambda = 1$, are always above the minorant curve

$$m(\tau, 1/2) = 0.25e^{\tau/2}. \tag{4.50}$$

A graphical representation of this exponentially growing curve, and one example from half of the realizations of the stochastically growing GBM that are always larger than the given minorant curve, are depicted in Fig. 4.4.

Substituting $V = \lambda$ in (4.48) and (4.49), we obtain the cdf

$$Q_-(s, s_0, \lambda) = \left(\frac{s}{s_0}\right)^\lambda, \quad 0 < s < s_0, \quad \lambda > 0, \tag{4.51}$$

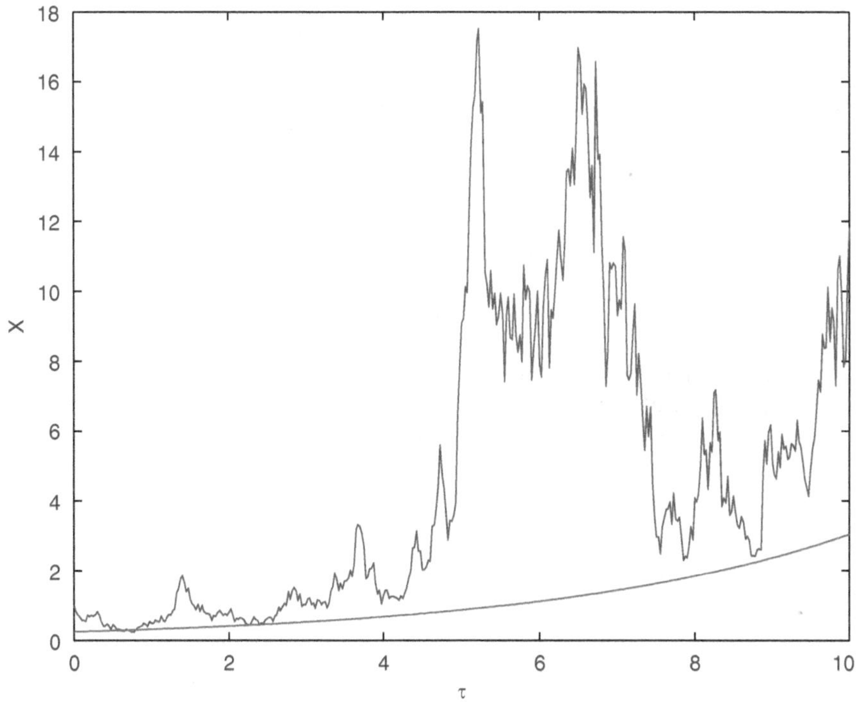

Fig. 4.4 Plot of the exponentially growing minorant curve (4.50), and one example among half of the realizations of the stochastically growing GBM (2.11) corresponding to $\lambda = 1$, which are always larger than the mentioned minorant curve

of the minimum values of the process $S(t)$ defined by (2.10):

$$S_- := \min_{t \in [0,\infty)} S(t). \tag{4.52}$$

Interpreting $Q_-(s, s_0, \lambda)$ as the probability that the diffusion process $S(t)$ (2.10) crosses a given level s, and taking into account that for the case under inspection ($\lambda > 0$), the diffusion process $S(t)$ necessarily crosses any level $s > s_0$, we finally obtain

$$Q_-(s, s_0, \lambda) = \begin{cases} \left(\dfrac{s}{s_0}\right)^\lambda, & 0 < s < s_0, \\ 1, & s > s_0, \end{cases} \qquad \lambda > 0. \tag{4.53}$$

4.7 Quantile Curves

It is clear from the above discussion that the properties of stochastically growing and stochastically decaying processes provide important insights for understanding the dependence on s of the mean density $g(s,t)$ of firm sizes. This motivates an investigation of more detailed properties of the realizations of stochastically growing and stochastically decaying processes. In this section, we consider some transparent characteristic of stochastic processes, embodied by the notion of *quantile curves*, which provide additional insight on the behavior of realizations of the GBM $S(t)$ defined by (2.9).

Let us first introduce the notion of quantile curves.

Definition 4.7.1. Given a diffusion process $\{S(t)\}_{t \in \mathbb{R}_+}$, the mapping $S_q(t)\colon \mathbb{R}_+ \mapsto \mathbb{R}$ is called a *quantile curve* at probability level q if

$$\forall t, \quad \Pr\left[S(t) > S_q(t)\right] = q. \tag{4.54}$$

In order to understand the underpinning of this notion of quantile curves for an arbitrary stochastic process $S(t)$, consider again its complementary cdf $\bar{F}(s;t)$ defined by (4.1). After integrating it over some time interval $[t_1, t_2]$, we obtain

$$\int_{t_1}^{t_2} \bar{F}(s;t)dt = E\left[\int_{t_1}^{t_2} \mathbf{1}(S(t) - s)\, dt\right]. \tag{4.55}$$

The integral

$$\tilde{T}(s, t_1, t_2) := \int_{t_1}^{t_2} \mathbf{1}(S(t) - s)\, dt \tag{4.56}$$

in the r.h.s. of (4.55) has the simple geometric meaning of being the total random duration $\tilde{T}(s, t_1, t_2)$ of the time intervals during which a given realization of the stochastic process $S(t)$ remains above the level s within the total interval $t \in [t_1, t_2]$.

Correspondingly, both parts of equality (4.55) are equal to the expected value of the total duration of the mentioned time intervals

$$\int_{t_1}^{t_2} \bar{F}(s;t)dt = T(s,t_1,t_2), \tag{4.57}$$

where

$$T(s,t_1,t_2) := E[\tilde{T}(s,t_1,t_2)]. \tag{4.58}$$

Consider now some function $S_q(t)$, which, for any time t, satisfies the equation

$$\bar{F}[S_q(t);t] = q, \tag{4.59}$$

for some given value $q \in (0,1)$. Integrating (4.59) over the interval $t \in [t_1,t_2]$ yields

$$\int_{t_1}^{t_2} \bar{F}(S_q(t);t)dt = q \cdot (t_2 - t_1). \tag{4.60}$$

The integral in the l.h.s. of (4.60) is the expected value $T_q(x)$ of the total duration of time intervals, belonging to the global interval $[t_1,t_2]$, within which the realizations of the stochastic process $S(t)$ remain above the curve $S_q(t)$. Thus, one may rewrite the last relation in the form

$$T_q(s) := \int_{t_1}^{t_2} \bar{F}(S_q(t);t)dt = q \cdot (t_2 - t_1). \tag{4.61}$$

In other words, the function $S_q(t)$ is such that, for any time interval (t_1,t_2), the fraction of expected time, during which the realization of the stochastic process $S(t)$ remains above level s, is equal to q. It is natural to refer to such functions as quantile curves.

Let us determine the quantile curves for the GBM $S(t)$ given by (2.9), whose distribution $f(s;t)$ is log-normal

$$f(s;t) = \frac{1}{s\sqrt{2\pi t}\,b} \exp\left[-\frac{1}{2b^2 t}\left(\ln\left(\frac{s}{s_0}\right) - ct\right)^2\right]. \tag{4.62}$$

Correspondingly, its complementary cdf is equal to

$$\bar{F}(s;t) = \int_s^\infty f(s',t)ds' = \frac{1}{2}\left(1 + \text{erf}\left[\frac{1}{\sqrt{2t}\,b}\ln\left(\frac{s_0}{s}\,e^{ct}\right)\right]\right).$$

Thus, the quantile curves of the GBM $S(t)$ are given by

$$S_q(t) = s_0\, e^{ct - p(q)\sqrt{2t}\,b} = s_0\, e^{\lambda\tau - 2p(q)\sqrt{\tau}}, \qquad \tau = \frac{b^2 t}{2}, \quad \lambda = \frac{2c}{b^2}, \tag{4.63}$$

where $p(q)$ is the solution to the equation

$$\mathrm{erf}(p) = 2q - 1.$$

Notice that, if q is close to 1, then, at almost all time, the realizations of the stochastic process $S(t)$ remain above the quantile curve $S_q(t)$, as expected. In this sense, one can say that the quantile curves for $q \to 1$ constitute weak versions of the minorant curves discussed previously. Analogously, for q close to zero, the quantile curves are weak versions of the majorant curves. Plots of some quantile curves for stochastically growing ($c > 0$) and stochastically decaying ($c < 0$) geometric Brownian motions are depicted in Figs. 4.5 and 4.6.

Figure 4.7 presents a quantile curve $S_q(t)$ for $\lambda = -1$, $q = 0.1$, together with a typical realization of the corresponding stochastically decaying process $S(t)$, such that the expected fraction of time intervals, for which $S(t) > S_q(t)$, is equal to 0.1.

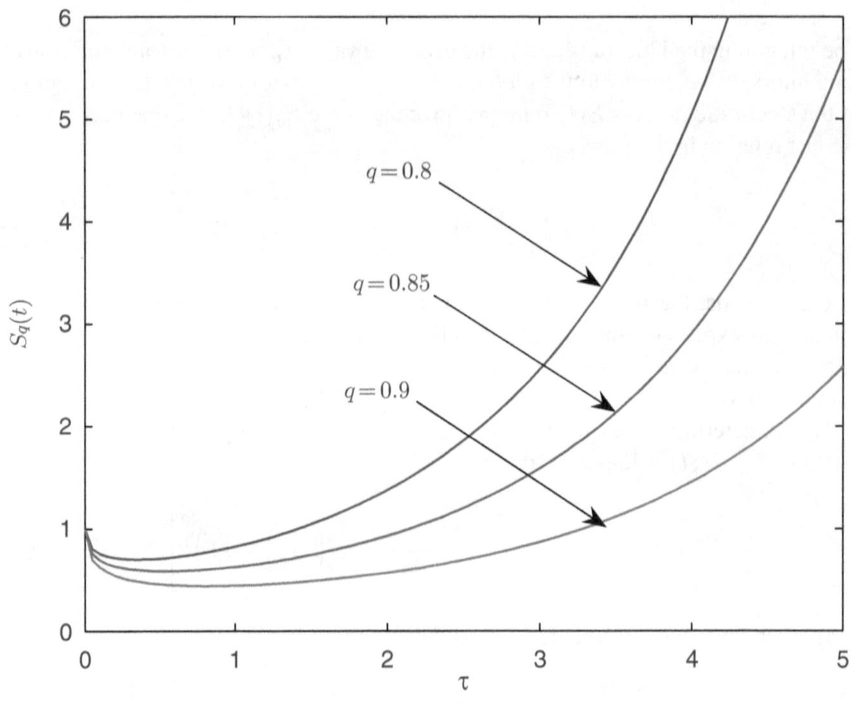

Fig. 4.5 Plots of "minorant" quantile curves for stochastically growing ($c = b^2/2$) GBM $S(t)$, for $s_0 = 1$ and for $q = 0.8, 0.85, 0.9$. These large values of q imply that the realizations of stochastically growing GBM $S(t)$ spend most of their time above these growing quantile curves, at least for large times. A typical realization of the corresponding stochastically growing process $S(t)$ is shown

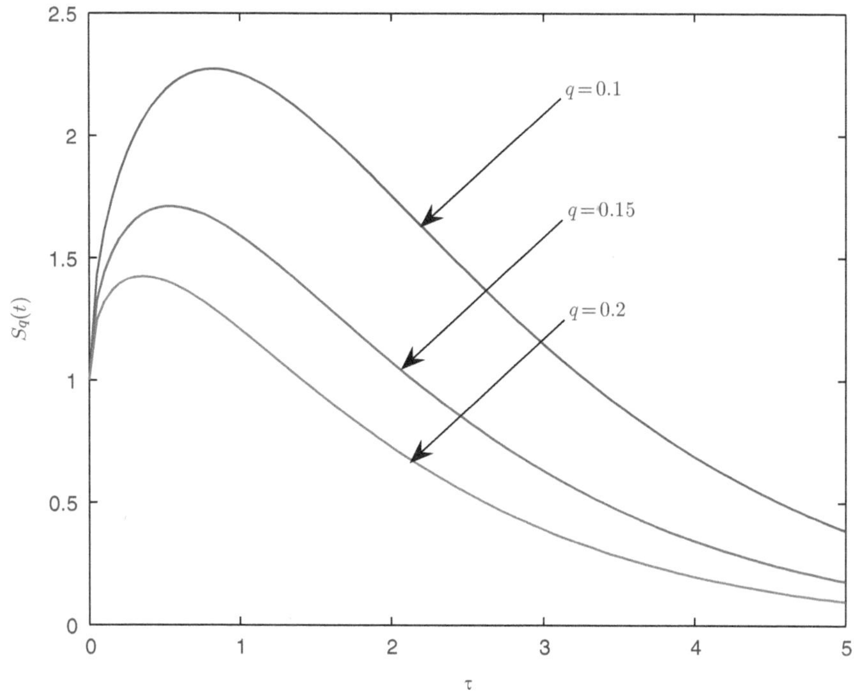

Fig. 4.6 Plots of "majorant" quantile curves for stochastically decaying ($c = -b^2/2$) GBM $S(t)$, for $q = 0.1, 0.15, 0.2$. Realizations of the stochastically decaying GBM $S(t)$ spend most of their time underneath these decaying quantile curves at large times. A typical realization of the corresponding stochastically decaying process $S(t)$ is shown

4.8 Geometric Explanation of the Steady-State Density of a Firm's Asset Value

The properties of GBMs discussed above allow us to present new transparent geometrical interpretations of the expressions (3.24), (3.25) for the steady-state mean density $g(s)$ of a firm's asset value given by (3.18). In this goal, suppose for a while that a firm's asset value $S(t)$ obeys the deterministic initial value problem

$$\frac{dS(t)}{dt} = c\left[S(t)\right], \qquad S(0) = s_0. \tag{4.64}$$

In this case, the pdf of the firm's asset value is singular

$$f(s;t) = \delta\left[S(t) - s\right]. \tag{4.65}$$

Below we will suppose for simplicity that $c(s)$ is a smooth function, strictly negative for $s \leqslant s_0$ or strictly positive for $s \geqslant s_0$.

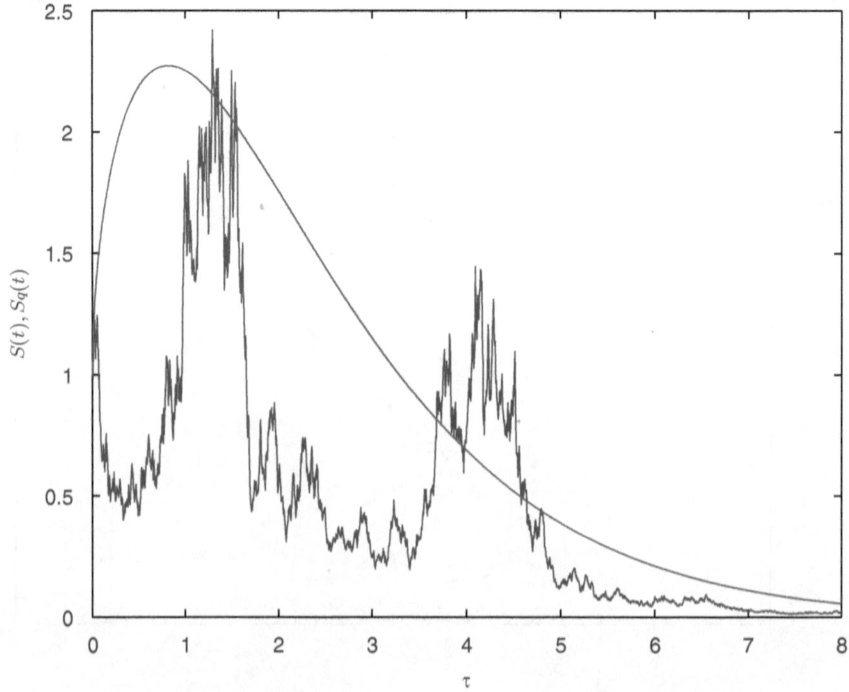

Fig. 4.7 Plots of the quantile curve $S_q(t)$ for $\lambda = -1$, $q = 0.1$, and a typical realization of the corresponding stochastically decaying process $S(t)$

Substituting (4.65) into the r.h.s. of expression (3.18) for the steady-state density of firm's asset values, and using both (4.64) and the following property of delta functions

$$\delta\left[S(t) - s\right] = \left|\frac{dS(t_*)}{dt_*}\right|^{-1} \delta(t - t_*), \tag{4.66}$$

where t_* is the root of the equation $S(t) = s$, we obtain

$$g_0(s) = \frac{\nu}{c(s)} \mathbf{1}(s - s_0). \tag{4.67}$$

We introduced the subscript "0" to distinguish the *deterministic* growth steady-state density $g_0(s)$ from the *random* growth steady-state mean density $g(s)$.

We supposed above that $c(s) > 0$ for any $s > s_0$. Similarly, if $c(s) < 0$ for any $s < s_0$, then

$$g_0(s) = \frac{\nu}{|c(s)|} \mathbf{1}(s_0 - s). \tag{4.68}$$

The geometrical meaning of these two relations (4.67) and (4.68) is obvious: the greater the absolute value of the velocity $c(s)$ when crossing some given level s, the smaller the steady-state density $g_0(s)$. In other words, the steady-state density

$g_0(s)$ is the inverse of the absolute value of the velocity at crossing level s. Looked at in another way, the unit step functions in (4.67) and (4.68) take into account that, if $c > 0$, then $S(t)$ never crosses any level $s < s_0$ and vice versa. Substituting $c(s) = c \cdot s$ in these expressions, we obtain in particular that

$$g_0(s) = \frac{\nu}{|c|s} \tag{4.69}$$

for

$$\begin{cases} s > s_0 & \text{if} \quad c > 0 \quad (\lambda > 0), \\ 0 < s < s_0 & \text{if} \quad c < 0 \quad (\lambda < 0). \end{cases} \tag{4.70}$$

Amazingly, $g_0(s)$ given by (4.69) coincides with the corresponding parts of the mean densities (3.24), (3.25). One can interpret this fact in the sense that, within the intervals defined in (4.70), the stochastic component of the asset value's $S(t)$ behavior does not influence the shape of the mean density $g(s)$.

In contrast, when the deterministic densities (4.67) and (4.68) are equal to zero, the stochastic component of $S(t)$ plays the crucial role. Let us determine it for the case when the density corresponds to the stochastically decaying GBM of firm's asset values given by (3.24). The density $g(s)$ is then transformed into

$$g(s) = \frac{\nu}{|c|s} Q_+(s, s_0, \lambda), \qquad \lambda < 0, \tag{4.71}$$

where the unit step function $\mathbf{1}(s_0 - s)$ in the r.h.s. of expression (4.68), which holds in the case where the deterministic drift dominates, is replaced by the probability $Q_+(s, s_0, \lambda)$ given by (4.46) that the diffusion process $S(t)$ crosses a given level s.

Analogously, expression (3.25) may be rewritten in the geometrically transparent form as

$$g(s) = \frac{\nu}{|c|s} Q_-(s, s_0, \lambda), \tag{4.72}$$

where $Q_-(s, s_0, \lambda)$ is the probability given by (4.53) that a stochastically growing process $S(t)$ defined by (2.10) crosses a level $s > 0$.

Therefore, both steady-state mean densities (3.24) and (3.25) have a clear structure: they are equal to the product of two factors. The first one is the "deterministic" density $g_0(s)$ (4.69), taking into account the regular drift of the GBM $S(t)$ (2.10). The second one is the probability $Q_\pm(s, \delta)$ that the diffusion process $S(t)$ crosses a given level s. For the case of balanced GBMs $S(t)$, and for $s > s_0$, both of the aforementioned factors $g_0(s)$ and $Q_+(s, s_0, \lambda)$ are proportional to $1/s$, thus contributing equally to Zipf's law $g(s) \sim 1/s^2$.

Having developed an intimate understanding of Zipf's law and its deviations in the presence of a flow of firm births, the next chapter adds a key ingredient, namely the fact that firms do not live forever but may disappear when their sizes become smaller than a minimum viable threshold. Chapter 7 will examine the impact of an addition channel of firm demises due to abrupt shocks, such as those resulting from operational risks.

Chapter 5
Exit or "Death" of Firms

5.1 Empirical Evidence and Previous Works on the Exit of Firms

Many theoretical models neglect the possibility that firms disappear. However, firms do not continuously grow. They undergo transient periods of decay that are sometimes persistent and then surrender to their decline which may ultimately lead to their exit from business.

Referring as in Chap. 2 to Bonaccorsi Di Patti and Dell'Ariccia (2004), the yearly rate of death of Italian firms is, on average, equal to 5.7% with a maximum of about 20% for some specific industry branches. Knaup (2005) examined the business survival characteristics of all establishments that started in the United States in the late 1990s when the boom of much of that decade was not yet showing signs of weakness, and found that, if 85% of firms survive more than one year, only 45% survive more than four years. Brixy and Grotz (2007) analyzed the factors that influence regional birth and survival rates of new firms for 74 West German regions over a 10-year period. They documented significant regional factors as well as variability in time: the 5-year survival rate fluctuates between 45% and 51% over the period from 1983 to 1992.

Bartelsman et al. (2003) confirmed that a large number of firms enter and exit most markets every year in a group of ten OECD countries: data covering the first part of the 1990s show the firm turnover rate (entry plus exit rates) to be between 15% and 20% in the business sectors of most countries: i.e., a fifth of firms are either recent entrants, or will close down within the year. In the CRSP database of 26,800 firms quoted on the North American markets covering the period from January 1926 to December 2006 (http://www.crsp.com/), Daniel et al. (2008) found that 25% of names disappear after 3.3 years, 75% of names disappear after 14 years and 95% of names disappeared after 34 years.[1] Using an exhaustive list of Japanese bankruptcy in 1997, Fujiwara (2004) showed that the distribution of total liabilities

[1] Disappearance from the CRSP database is not uniquely associated with the exit of a firm in the sense used in this book, but may result from merger and acquisition, or delisting. Section 10.1 briefly discuss a model of merger and acquisition which generalizes the GBM model.

A. Saichev et al., *Theory of Zipf's Law and Beyond*,
Lecture Notes in Economics and Mathematical Systems 632,
DOI 10.1007/978-3-642-02946-2_5, © Springer-Verlag Berlin Heidelberg 2010

of bankrupted firms in the high debt range also obeys Zipf's law. While it has been established that a first-order characterization for firm death involves lower failure rates for larger firms (Dunne et al., 1988, 1989; Bartelsman et al., 2003), Bartelsman et al. (2003) also state that for sufficiently old firms, there seems to be no difference in the firm failure rate across size categories.

To the extent that the empirical literature documents a sizable exit at all size categories, we suggest that it is timely to study different models with both firm exit at a size lower bound and due to a size-independent hazard rate.

Simon (1960) as well as Steindl (1965) have considered this stylized fact and provide a generalization of Simon (1955) where the decline of a firm and ultimately its exit is accounted for when its size reaches zero. Both Simon's and Steindl's models involve a discrete scale of firm sizes since, for each time interval, the growth of a firm is restricted to one size unit up or down. In Simon (1960)'s model, the rate of firm's exits exactly compensates for the flow of firm's births so that the economy is stationary and the steady-state distribution of firm sizes exhibit the same upper tail behavior as in Simon (1955). In contrast, Steindl (1965) includes births and deaths but within an industry with a growing number of firms. A steady-state distribution is obtained whose tail follows a power law with an exponent which depends on the net entry rate of new firms and on the average growth rate of incumbent firms. Zipf's law is only recovered in the limit where the net entry rate of new firms goes to zero.

Both models rely on the existence of a minimum size below which a firm runs out of business. This hypothesis corresponds to the existence of a minimum efficient size below which a firm cannot operate, as is well established in economic theory. However, there may be in general more than one minimum size as the exit (death) level of a firm has no reason to be equal to the size of a firm at birth. In the afore mentioned models, these two sizes are assumed to be equal, while there is a priori no reason for such an assumption. Moreover, this minimum size is taken to be the same for all firms. Again, this assumption is counterfactual and needs to be relaxed.

In addition to the exit of a firm resulting from its value decreasing below a certain level, it sometimes happens that a firm encounters financial troubles while its asset value is still fairly high. Recent striking examples are Enron Corp. and Worldcom whose market capitalization were still high (actually the result of inflated total asset value of about $11 billion for Worldcom and probably much higher for Enron) when they bankrupted. As a consequence, it is also necessary to account for the disappearance of firms whose total asset values are still far from the lower limit determining the smallest efficient size. Gabaix (1999) considers an analogous situation (at least from a mathematical perspective) and suggests that it may have an important impact on the shape of the distribution of firm sizes.

The above empirical facts show that one should enrich the model describing the stochastic behavior of any given asset value to take into account the possible death of firms. This chapter is devoted to discussing various aspects of the dynamics of firm's asset values modeled by the stochastically decaying GBM ($\lambda < 0$), in the presence of death that occurs when a firm asset value becomes smaller than a given pre-defined level s_1.

5.2 Life-Span Above a Given Level

Before implementing the actual exit of firms whose sizes become smaller than s_1, one can introduce the concept of a finite life-span in an otherwise infinitely lived decaying GBM ($\lambda < 0$) defined by (2.9), as being the time it takes for the firm to see its size shrink below a small level $s_1 > 0$.

While the statistical average of the total number of firms, $\int_0^\infty g(s)ds$, where $g(s)$ is the steady-state mean density given by Proposition 3.4.1 is infinite, it turns out that the mean number of firms whose asset values are larger than any strictly positive level s_1 is finite. Indeed, using the mean steady-state density (3.24) corresponding to stochastically decaying asset values ($\lambda < 0$), we obtain

$$G(s_1) = \int_{s_1}^{\infty} g(s)ds = \frac{\nu}{|c\lambda|} + \frac{\nu}{|c|} \ln\left(\frac{s_0}{s_1}\right), \qquad 0 < s_1 < s_0. \tag{5.1}$$

The cumulative density $G(s_1)$ has the transparent meaning of being equal to

$$G(s_1) = \nu T(s_1), \tag{5.2}$$

where $T(s_1)$ is the mean duration of the firm life above level s_1. Thus, for the mean density of firm sizes given by (3.24), the mean life duration above level s_1 is finite:

$$T(s_1) = \int_{s_1}^{\infty} g(s)ds = \frac{1}{|c\lambda|} + \frac{1}{|c|} \ln\left(\frac{s_0}{s_1}\right) < \infty. \tag{5.3}$$

In order to more accurately estimate the statistical properties of life durations above some arbitrary level s_1, we formulate the following definition.

Definition 5.2.1. Let some firm be born at time $t_0 = 0$ and let $\{S(t)\}_{t\in\mathbb{R}_+}$ be the diffusion process describing the evolution of its size. The life duration $t_d(s_1)$ above level $s_1 > 0$ is the last instant when the firm size exceeds the given level s_1.

$$t_d\,(s_1) := \sup\left\{t \in \mathbb{R}_+ \mid \forall t' \leq t,\; S(t') \geq s_1\right\}. \tag{5.4}$$

Correspondingly, we will call instant t_d the *instant of natural death* of the firm under inspection.

When applied to Gabaix's context of city sizes, it makes sense to choose $s_1 = s_0$, interpreting s_0 as a minimal size above which some settlement gains the status of city, while below s_0 it loses that status. In accordance with the above definition, any city, even if it has been very large in the past, will never be qualified as city for any $t > t_d$. Figure 5.1 illustrates the definition of a firm's (city's) natural death.

In the context of firms, the above definition of the life duration t_d can be interpreted as the time beyond which a procedure such as chapter 11 of the United States Bankruptcy Code will not revive the firm. When a troubled business is unable to service its debt or pay its creditors, usually because its assets become too low, the firm

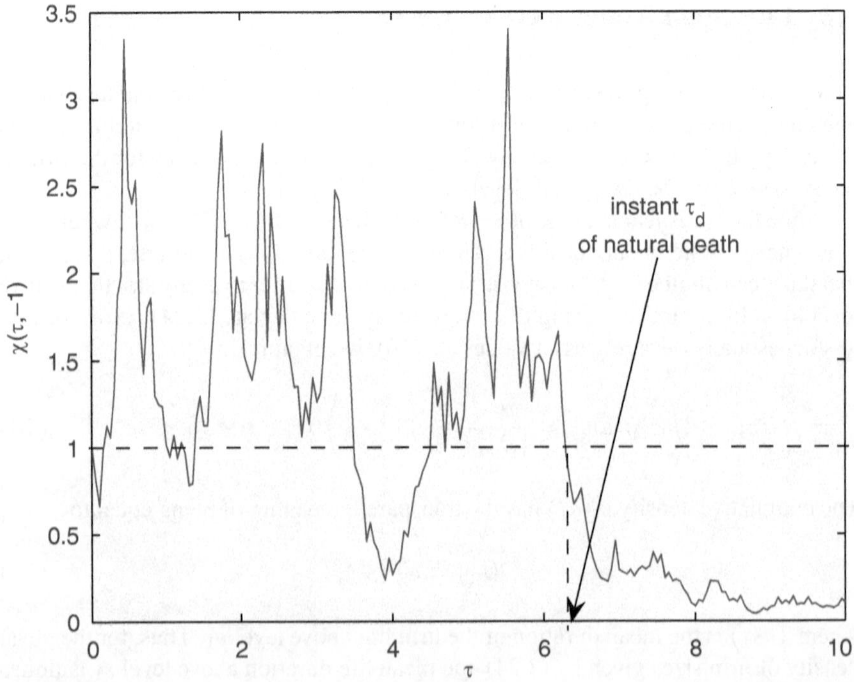

Fig. 5.1 A plot of one realization of the balanced GBM $\chi(\tau, -1)$ defined by (2.17), illustrating the notions of life duration above a given level and the corresponding instant of natural death. Here the level of birth and death are identical and are equal to 1. The *arrow* points out the dimensionless value of the instant of natural death $\tau_d = t_d/t_b$

or its creditors can file with a federal bankruptcy court for protection. One of these procedures is the chapter 11 filing, which is usually an attempt to stay in business while a bankruptcy court supervises the reorganization of the company's contractual and debt obligations. The firm's asset value depicted in Fig. 5.1 exhibits several intervals, the largest one around the reduced time $\tau \approx 4$, during which a chapter 11 filing would be needed to ensure the survival the firm.

5.3 Distribution of Firm's Life Durations Above a Survival Level

Let us prove that, for stochastically decaying GBMs, the life duration $t_d(s_1)$ above any level $s_1 > 0$ is almost surely finite. This is equivalent to the fact that the complementary cdf $\bar{F}_d(t)$ of the random life durations, defined by

$$\bar{F}_d(t) := \Pr\{t_d(s_1) > t\} \tag{5.5}$$

tends to zero as $t \to \infty$.

Using the Markovian properties of the firm's asset value $S(t)$ defined in (2.9), one can express the probability that the random instant $t_d(s_1)$ of the firm's death is smaller than a given t in the form

$$\Pr\{t_d(s_1) \leq t\} = F_d(t) = \int_0^{s_1} f(u; s_0, t) P_+(s_1, u, \lambda) du, \tag{5.6}$$

where $P_+(s_1, u, \lambda)$ given by (4.41) is the probability that if, at a given time t, the diffusion process $S(t)$ defined by (2.9) has the value $u < s_1$, then for any $t' > t$ the process $S(t')$ will not exceed the level s_1. Calculating the integral (5.6), one gets

$$\begin{aligned} \bar{F}_d(t) = 1 - \frac{1}{2} \Bigg(& \operatorname{erfc}\left[\frac{1}{2\sqrt{t/t_b}} \ln\left(\frac{s_0}{s_1} e^{\lambda \frac{t}{t_b}}\right)\right] \\ & - \left(\frac{s_1}{s_0}\right)^{\lambda} \operatorname{erfc}\left[\frac{1}{2\sqrt{t/t_b}} \ln\left(\frac{s_0}{s_1} e^{-\lambda \frac{t}{t_b}}\right)\right]\Bigg). \end{aligned} \tag{5.7}$$

It is easy to show from expression (5.7) that, if the GBM $S(t)$ is stochastically decaying ($\lambda < 0$), then $\bar{F}_d(t)$ converges to zero exponentially fast as $t \to \infty$.

Corresponding to (5.7), the pdf of firm's dimensionless life duration $\tau_d(s_1) = t_d(s_1)/t_b$ above a given level s_1 is equal to

$$f_d(\tau) = \frac{|\lambda|}{2\sqrt{\pi\tau}} \exp\left[-\frac{1}{4\tau}\left(\ln \frac{s_0}{s_1} + \lambda\tau\right)^2\right], \quad \lambda < 0. \tag{5.8}$$

In particular, if the initial size coincides with the level of death ($s_1 = s_0$), then one gets

$$f_d(\tau) = \frac{|\lambda|}{2\sqrt{\pi\tau}} \exp\left(-\frac{\lambda^2}{4}\tau\right). \tag{5.9}$$

Plots of the pdf given by (5.8) of life durations for different ratios of s_0/s_1 for the case of the balanced Gibrat's law ($\lambda = -1$) are depicted in Fig. 5.2.

5.4 Killing of Firms upon First Reaching a Given Asset Level from Above

In the previous section, we have proved that, if one applies the status of firm only to those which have asset values greater than or equal to some qualified amount $s_1 > 0$ then, if $\lambda < 0$ and in particular for the balanced Gibrat's law model ($\lambda = -1$), each firm almost surely has a finite life duration $t_d(s_1) < \infty$. However, as we did not specified any dismantling or disrupting process, the firm actually still exists, possessing an asset value smaller than s_1 ($S(t) < s_1$ for any $t > t_d$). One might

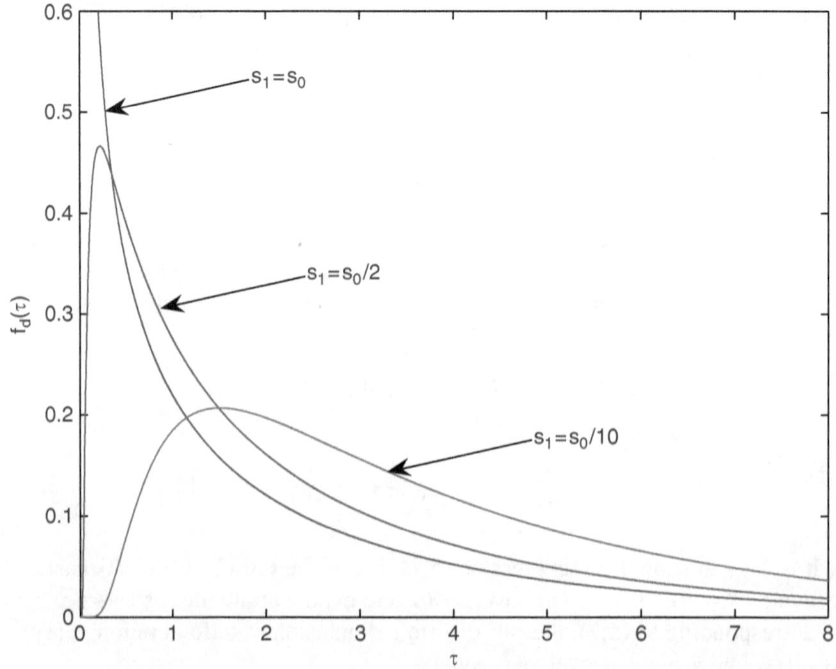

Fig. 5.2 Plots of the pdf (5.8) of life durations, defined as the last instant when the firm asset value was above a given level s_1 for the case of balanced Gibrat's law and for different ratios of s_0 and s_1: $s_1 = s_0$; $s_1 = s_0/2$ and $s_1 = s_0/10$

say that the firm survives with an insignificant asset value, never exceeding again the marginal level s_1.

In contrast, in this and the following sections, we consider another realistic situation such that

Assumption 4 *At the first crossing of the given level s_1 which is considered to be insufficient for further activity, the firm is dismantled.*

Figure 5.3 makes clear the distinction between the killing process of the firm when its value becomes smaller than some threshold s_1 for the first time, and the demise of the firm occurring in general at a later time beyond which the firm, which is still allowed to survive, never reaches again the level s_1.

In order to take into account the killing of a firm when it first crosses level s_1, we make use of the statistical description of the truncated Wiener process with drift $Y(t, c, b)$ (2.6), which is killed when first crossing a given level $-y_1$, where

$$y_1 := \ln\left(\frac{s_0}{s_1}\right), \qquad 0 < s_1 < s_0. \tag{5.10}$$

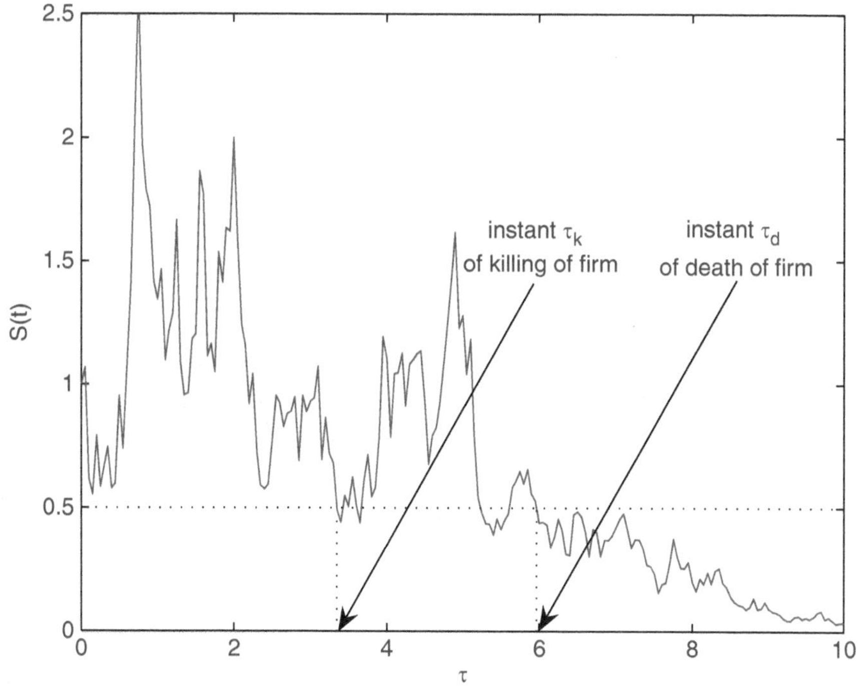

Fig. 5.3 An illustration comparing killing and natural death instants at a given level s_1 (taken to be half of the initial value $s_0 = 1$) for the same realization of the balanced GBM (2.9). The "killing" time τ_k corresponds to the first crossing of the level s_1 from above. The instant τ_d is the last time that the firm was found above level s_1

It is known that the pdf $h(y;t)$ of such truncated Wiener process with drift satisfies the diffusion equation

$$\frac{\partial h(y;t)}{\partial t} + c\frac{\partial h(y,t)}{\partial y} = \frac{b^2}{2}\frac{\partial^2 h(y,t)}{\partial y^2} \tag{5.11}$$

with the following initial and absorbing boundary conditions

$$h(y;t=0) = \delta(y), \qquad h(y=-y_1;t>0) = 0. \tag{5.12}$$

Solving the initial-boundary problem (5.11), (5.12), by the reflection method (Morse and Feshbach, 1953), one gets

Proposition 5.4.1. *The pdf of the truncated Wiener process $Y(t,c,b)$, solution to the mixed initial-boundary problem (5.11)–(5.12) is*

$$h(y;t) = \phi(y;t) - e^{-\lambda y_1}\,\phi(y+2y_1;t), \tag{5.13}$$

where the Gaussian distribution $\phi(y;t)$ is described by expression (2.8).

In turn, given Proposition 5.4.1, one can obtain the pdf $f(s; s_0, t)$ of the firm's asset value $S(t)$, taking into account the killing of the firm upon its asset value first touching the level s_1, using the formula

$$f(s; s_0, t) = \frac{1}{s} h \left[\ln \left(\frac{s}{s_0} \right); t \right], \tag{5.14}$$

which follows from the functional relation (2.10) with (2.11) between the Wiener process with drift $Y(t, c, b)$ and the GBM $S(t)$.

To better realize the statistical meaning of the pdf $f(s; s_0, t)$ given by (5.14), consider its transparent interpretation. Imagine that there were $N \gg 1$ firms born at the same instant $t_0 = 0$. Then, $N f(s; s_0, t)$ describes the mean density of the firm's asset values, taking into account their killing upon first touching the level s_1. In particular, the integral

$$\int_{s_1}^{\infty} f(s; s_0, t) ds = \int_{-y_1}^{\infty} h(y; t) dy, \tag{5.15}$$

is equal to the fraction of the original N firms surviving until the current instant t.

5.5 Life-Span of Finitely Living Firms

Note that the fraction of surviving firms given by (5.15) has a clear probabilistic interpretation. It provides the complementary cdf of life durations of the firms

$$\bar{F}_k(t) := \Pr\{t_k(s_1) > t\}, \tag{5.16}$$

where $t_k(s_1)$ is the random instant of a firm's exit. Recall that Fig. 5.3 makes clear the difference between the random instant $t_k(s_1)$ of a firm's killing upon first touching the level s_1, and the instant $t_d(s_1)$ of firm's natural death at the same level s_1.

Substituting expression (5.13) into the last integral of equality (5.15), and calculating the integral analytically, we obtain (with the notation $\tau = t/t_b$)

$$\begin{aligned} \bar{F}_k(\tau) = \frac{1}{2} \Bigg(& \operatorname{erfc} \left[\frac{1}{2\sqrt{\tau}} \ln \left(\frac{s_1}{s_0} e^{-\lambda\tau} \right) \right] \\ & - \left(\frac{s_1}{s_0} \right)^{\lambda} \operatorname{erfc} \left[\frac{1}{2\sqrt{\tau}} \ln \left(\frac{s_0}{s_1} e^{-\lambda\tau} \right) \right] \Bigg). \end{aligned} \tag{5.17}$$

For large $\tau = t/t_b$, expression (5.17) reduces to the accurate asymptotic

$$\bar{F}_k(\tau) = e^{-\lambda^2 \tau/4} \frac{2}{\lambda^2 \tau \sqrt{\pi\tau}} \left(\frac{s_0}{s_1} \right)^{-\lambda/2} \ln \left(\frac{s_0}{s_1} \right). \tag{5.18}$$

The corresponding pdf of the dimensionless life-span of finitely living firms is

$$f_k(\tau) = \frac{\ln\left(\frac{s_0}{s_1}\right)}{2\tau\sqrt{\pi\tau}} \exp\left[-\frac{1}{4\tau}\left(\ln\frac{s_0}{s_1} + \lambda\tau\right)^2\right]. \tag{5.19}$$

Note that $f_k(\tau)$ is related to the pdf $f_d(t)$ given by (5.8) of life durations above a given level s_1 by the following elegant relation

$$f_d(\tau) = \frac{\tau}{E[\tau_k]} f_k(\tau), \tag{5.20}$$

where

$$E[\tau_k] = \int_0^\infty f_k(\tau)d\tau = -\frac{1}{\lambda}\ln\left(\frac{s_0}{s_1}\right) \qquad (\lambda < 0) \tag{5.21}$$

is the mean life duration of killed firms.

5.6 Influence of Firm's Death on the Balance Condition

We have introduced above the natural balance condition (Definition 2.5.1), which one may interpret as a consequence of the invariance of the statistical average of the overall asset value of firms established at the same original instant t_0. Because any firm in this model is statistically equivalent to any other one, the balance condition reduces to

$$E[S(t)] = \text{const.}, \tag{5.22}$$

where $S(t)$ is the current asset value of some firm which was born at instant $t_0 = 0$. Remember also that, if $S(t)$ obeys the pure Gibrat's law, such that the life of each firm is infinitely long, then equality (5.22) is valid if $\lambda = -1$.

One might assume that the inclusion of firm's death causes the balance condition to break, because the fraction of surviving firms out of firms which were born at the same original instant $t_0 = 0$ should decay with the growth of time t. It is worthwhile exploring the influence of firm's exits on the behavior of the firm's mean asset value E[S(t)], revealing whether the balance condition remains valid or not in the presence of firm's killing.

It is easy to calculate analytically the moments of the asset value $S(t)$ of a firm which has still not been killed at current instant t

$$E[S^m(t)] = \int_{s_1}^\infty s^m f(s; s_0, t)ds. \tag{5.23}$$

Substituting here the pdf (5.14) [with (5.13)], we obtain (in reduced time)

$$E[S^m(t)] = E_0[S^m(t)]\, K(\tau, s_0, s_1, \lambda), \tag{5.24}$$

where

$$E_0[S^m(t)] = s_0^m e^{m(m+\lambda)t} \tag{5.25}$$

is the corresponding moment of the pure GBM $S(t)$ given by (2.9), while the last factor

$$\begin{aligned} K(\tau, s_0, s_1, \lambda) = \frac{1}{2}\Bigg(&\operatorname{erfc}\left[\frac{1}{2\sqrt{\tau}}\ln\left(\frac{s_1}{s_0}e^{-(\lambda+2m)\tau}\right)\right] \\ &- \left(\frac{s_1}{s_0}\right)^{\lambda+2m} \operatorname{erfc}\left[\frac{1}{2\sqrt{\tau}}\ln\left(\frac{s_0}{s_1}e^{-(\lambda+2m)\tau}\right)\right]\Bigg) \end{aligned} \tag{5.26}$$

takes into account the influence of firm's deaths at a given level s_1. In particular, the average of the current firm's asset values, for the balanced case $\lambda = -1$, is equal to

$$\begin{aligned} E[S(t)] = \frac{1}{2}\Bigg(&s_0 \operatorname{erfc}\left[\frac{1}{2\sqrt{t}}\ln\left(\frac{s_1}{s_0}e^{-t}\right)\right] \\ &- s_1 \operatorname{erfc}\left[\frac{1}{2\sqrt{t}}\ln\left(\frac{s_0}{s_1}e^{-t}\right)\right]\Bigg). \end{aligned} \tag{5.27}$$

This expression predicts that, with the growth of the reduced time τ, the average of the firm's asset values (5.27) converges to a nonzero value

$$E[S(t=\infty)] = s_0 - s_1. \tag{5.28}$$

This convergence is illustrated in Fig. 5.4. This means in particular that, asymptotically, the killing of firms does not break the balance condition. In other words, the statistical average of the overall asset value of firms established at the same time is asymptotically constant.

Inasmuch as the current fraction $\bar{F}_k(\tau)$ of surviving firms tends to zero, the asymptotic balance condition (5.28) means that the average asset value of firms that have survived until current time τ,

$$E[S(t)|\text{surviving firms}] = E[S(t)]/\bar{F}_k(t), \tag{5.29}$$

grows exponentially as $\sim e^{\tau/4}$, for large $\tau \gg 1$ for the case $\lambda = -1$, as illustrated in Fig. 5.5.

5.7 Firm's Death Does Not Destroy Zipf's Law

It can be seen from the discussion of the previous section that it is possible to take into account the death of firms at a given level $s_1 > 0$ in the framework of stochastic growth models of firms asset values using Gibrat's law. We now consider in detail the mean density of firm's asset values taking into account the killing of firms upon first touching a given level s_1. It will be seen that firm's death does not destroy Zipf's law.

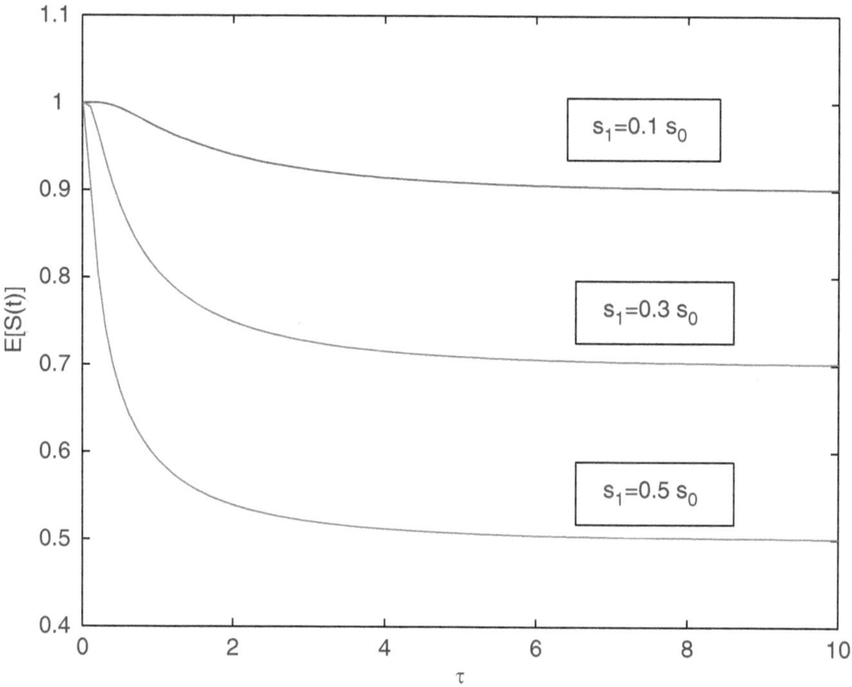

Fig. 5.4 Plots of the current mean asset value $S(t)$ (in case $s_0 = 1$), taking into account the killing of firms at a given level s_1, for the balanced condition $\lambda = -1$

Let us find the steady-state mean density of firms killed upon first touching level s_1 $(0 < s_1 < s_0)$ according to Definition 3.3.1. The proof of the existence of the steady-state mean density in the presence of the exit level (Assumption 4) is the same as in its absence (see the derivation of Proposition 3.3.1). It is therefore omitted. Thus, substituting (5.14)–(5.13) into (3.18), and supposing that $\lambda < 0$, we obtain

Proposition 5.7.1. *Under the assumptions of Proposition 3.3.1 and Assumption 4, a steady-state mean density of firm's size exists and is such that, for* $\lambda < 0$,

$$g(s) = \frac{\nu}{|c|s}\begin{cases} 1 - \left(\dfrac{s}{s_1}\right)^{\lambda}, & s_1 < s < s_0, \\ \left(\dfrac{s}{s_0}\right)^{\lambda}\left[1 - \left(\dfrac{s_0}{s_1}\right)^{\lambda}\right], & s > s_0, \end{cases} \quad . \tag{5.30}$$

It is worthwhile to compare this mean density with the mean density given in the absence of exit level by Proposition 3.4.1, corresponding to firms whose stochastic behavior is described by the non-truncated GBM given by (2.10), (2.11). It is seen that, due to firm's death at level s_1, the lower power tail $g(s) \sim s^{-1}$ inherent to

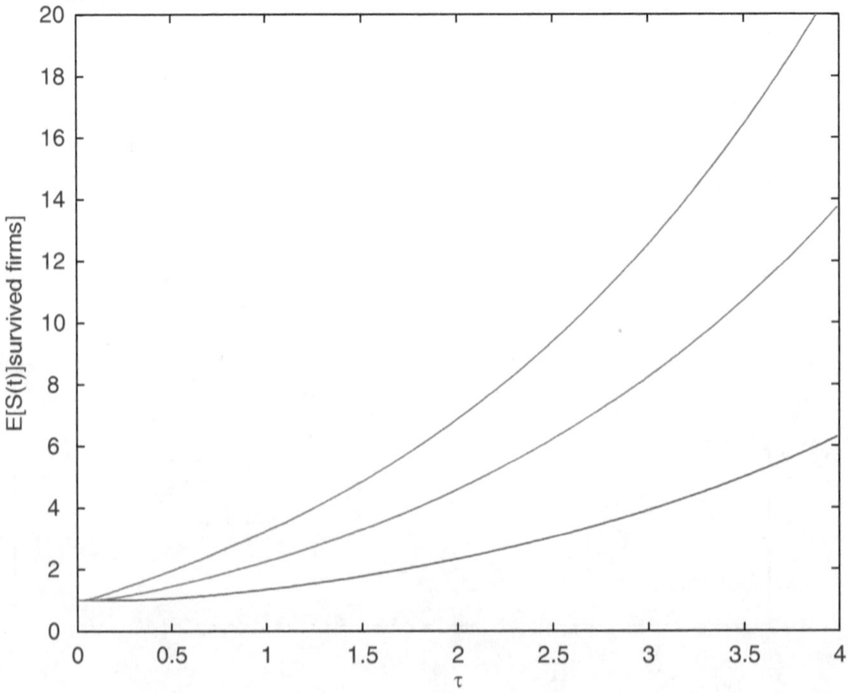

Fig. 5.5 Plots of the mean asset value of firms that have survived until the current time t (for $s_0 = 1$ and for the balanced case $\lambda = -1$). *From bottom to top* $s_1 = 0.1, 0.3, 0.5$

the steady-state density (3.24) is absent. On the other hand, for $s > s_0$ both densities (5.30) and (3.24) possess an upper power tail $g(s) \sim s^{\lambda-1}$ which corresponds to Zipf's law for the balanced case $\lambda = -1$. Figure 5.6 shows a log-log plot of the steady-state mean density (5.30) for the balanced case $\lambda = -1$, demonstrating the absence of a lower tail together with the presence of an upper Zipf's law tail. Unsurprisingly, the introduction of a mechanism for firm's killing only disturb the distribution for small s-values, while keeping unchanged the large s power law tail. It shows that detailed descriptions and understanding of the economic reasons underpinning the existence of a minimum firm size are not relevant for the explanation of Zipf's law, at least for large firm sizes.

5.8 Robustness Vis-a-vis the Randomness of Initial Firm's Sizes

The previous results are obtained for a fixed size s_0 at birth, which is identical for all firms. Let us now show that the asymptotic power law distribution of firm's sizes derived in Proposition 5.7.1 remains valid when taking into account that the

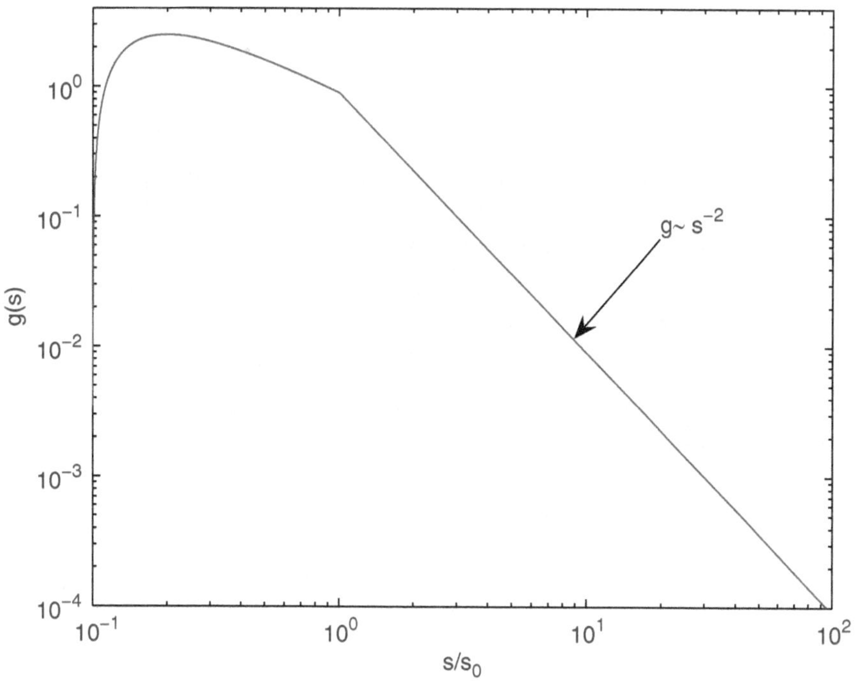

Fig. 5.6 Log-log plot of the steady-state mean density (5.30) for the balanced case $\lambda = -1$ and for $s_1 = s_0/10$. It is seen that the density satisfies Zipf's law for all $s > s_0$, while firm's death at level s_1 destroys the lower power tail for $s_1 < s < s_0$

initial sizes of firms are not identical, but are randomly drawn from some statistical distribution. We denote by f_{s_0} the density of the initial firm sizes. The density of a single firm $f(s,t)$ then solves

$$\frac{\partial f(s;t)}{\partial t} + \frac{\partial [a(s) f(s;t)]}{\partial s} = \frac{1}{2} \frac{\partial^2 [b^2(s) f(s;t)]}{\partial s^2} \tag{5.31}$$

with the initial condition

$$f(s;0) = f_{s_0}(s). \tag{5.32}$$

It is straightforward to show that

$$f(s;t) = \int f_0(s;t;s_0) \cdot f_{s_0}(s_0)\, ds_0 = \mathrm{E}_{s_0}\left[f_0(s;t;\tilde{s}_0) \right], \tag{5.33}$$

where $f_0(s;t;s_0)$ denotes the solution to (5.31) with the initial condition

$$f_0(s;0;s_0) = \delta(s - s_0). \tag{5.34}$$

The steady-state mean density of firm's size, with a random initial condition, reads

$$g(s) = \nu \int_0^\infty f(s; u)\, du = \nu \int_0^\infty \mathrm{E}_{s_0}\left[f_0(s; t; \tilde{s}_0)\right]\, du, \tag{5.35}$$

and, provided that we can change the order of the expectation and integration, we get

$$g(s) = \mathrm{E}_{s_0}\left[g_0(s; \tilde{s}_0)\right], \tag{5.36}$$

where

$$g_0(s; s_0) := \nu \int_0^\infty f_0(s; u; s_0)\, du \tag{5.37}$$

denotes the density of firm's size for a fixed initial condition s_0.

Using Proposition 5.7.1 together with (5.36), we get for $\lambda < 0$

$$\begin{aligned} g(s) &= \frac{\nu}{|c| s^{1-\lambda}} \left[\int_s^\infty f_{s_0}(s_0)\left(s^{-\lambda} - s_1^{-\lambda}\right) ds_0 + \int_{s_1}^s f_{s_0}(s_0)\left(s_0^{-\lambda} - s_1^{\lambda}\right) ds_0\right] \\ &= \frac{\nu}{|c| s^{1-\lambda}} \left[s^{-\lambda} \Pr\left(\tilde{s}_0 \geq s\right) - s_1^{-\lambda} + \int_{s_1}^s f_{s_0}(s_0) s_0^{-\lambda}\, ds_0\right] \\ &= \frac{\nu}{|c| s^{1-\lambda}} \left(\mathrm{E}\left[\tilde{s}_0^{-\lambda}\right] - s_1^{-\lambda} + o(s)\right), \quad \text{as } s \to \infty, \end{aligned} \tag{5.38}$$

provided that $\mathrm{E}\left[\tilde{s}_0^{-\lambda}\right] < \infty$. Thus

Proposition 5.8.1. *Under the assumptions of Proposition 5.7.1, the steady-state mean density of firm's size follows a power law with tail index $m = |\lambda|$ provided that the distribution of initial firm's size f_{s_0} admits a finite moment of order $|\lambda|$.*

Hence, as long as the above mentioned moment condition holds, i.e., as long as the distribution of initial firm's sizes is not too fat-tailed, the shape of the mean distribution of firm's sizes remains the same as found in absence of heterogeneity of initial firm sizes.

Chapter 6
Deviations from Gibrat's Law and Implications for Generalized Zipf's Laws

The introduction of a mechanism in which firms die introduces already a deviation from Gibrat's law for small s-values. Killing firms upon first touching the level $s_1 > 0$ actually means that the corresponding firm's asset values $S(t)$ do not obey strictly Gibrat's law of proportionate growth. Indeed, when $S(t)$ becomes close to s_1, the possibility of touching s_1 arises, and the rate $R(t, \Delta)$ given by (2.1) significantly depends on s_1.

In the present chapter, we will discuss in detail another general class of models in which the stochastic growth process deviates from Gibrat's law in different ways. Specifically, we will suppose that $S(t)$ is a diffusion process, obeying the stochastic equation

$$dS(t) = a[S(t)]dt + b[S(t)]dW(t), \qquad S(t=0) = s_0, \tag{6.1}$$

so that the corresponding pdf $f(s;t)$ satisfies the diffusion equation (2.39) and the initial condition (2.40). Recall that Gibrat's law of proportionate growth implies in particular that the coefficients $a(s)$ and $b(s)$ of the stochastic equation (6.1) are given by relations (2.41), i.e., are proportional to s. However, there is a wide and recent empirical literature, that suggests that Gibrat's law does not hold, in particular for small firms (Reid, 1992; Audretsch, 1995; Harhoff et al., 1998; Weiss, 1998; Audretsch et al., 1999; Almus and Nerlinger, 2000; Calvo, 2006) See however Lotti et al. (2003, 2007) for a dissenting view.

Thus, in order to get a more realistic description of the behavior of firm's asset values, we have to explore in detail the consequences of possible deviations of the coefficients $a(s)$ and $b(s)$ from direct proportionality to s, expressed by relations (2.41). We will quantitatively characterize the deviation from Gibrat's law of the stochastic process $S(t)$, satisfying the stochastic equation (6.1), by the ratios

$$A(s) := \frac{a(s)}{s}, \qquad B(s) := \frac{b(s)}{s}. \tag{6.2}$$

For instance, several groups have reported a nontrivial relationship between the size S of the firm and the variance $[B(S)]^2$ of its growth rate: $B(S) \sim 1/S^{\beta}$ with $\beta(S) \approx 0.2$ is an exponent that weakly depends on S (Stanley et al., 1996; Bottazzi

A. Saichev et al., *Theory of Zipf's Law and Beyond*,
Lecture Notes in Economics and Mathematical Systems 632,
DOI 10.1007/978-3-642-02946-2_6, © Springer-Verlag Berlin Heidelberg 2010

et al., 2001; Sutton, 2002; Podobnik et al., 2009). Similar results have been found also for economic variables such as exports, imports, foreign debt and the growth of the GDP (Lee et al., 1998; Podobnik et al., 2008). We note that such a dependence has the same qualitative monotonous decay as the semi-geometric Brownian process (6.78) corresponding to $B(s) = b(1 + \frac{\kappa}{s})$, which is analyzed in details below.

Obviously, the more the ratios (6.2) differ from constants, the stronger the deviations from Gibrat's law. As suggested by the empirical findings of Lotti et al. (2003, 2007), we will assume in some of the models of non-Gibrat's diffusion processes $S(t)$, that the larger $S(t)$ is, the better is the approximation provided by Gibrat's law. This means in particular that there is a non-zero finite limit

$$\lim_{s\to\infty} \frac{b(s)}{s} = b, \qquad 0 < b < \infty, \tag{6.3}$$

so that the asymptotic relation

$$b(s) \simeq b \cdot s, \qquad s \to \infty, \tag{6.4}$$

remains valid.

This assumption (6.3) differs from the prediction of models in which firms are considered to be aggregates of sub-units of uneven sizes (Fu et al., 2005; Pammolli et al., 2007; Buldyrev et al., 2007a,b; Riccabonia et al., 2008; Sakai and Watanabe, 2009). Assuming proportional growth in both the number of units in firms and their sizes, Fu et al. (2005) showed that the size-variance relationship is not a true power law with a single well-defined exponent but undergoes a slow crossover from $\beta = 0$ for $S \to 0$ to $\beta = 1/2$ for $S \to +\infty$ (Riccabonia et al., 2008).

The derivations of this chapter provide a general framework to treat the distribution of firm sizes, for arbitrary forms of $A(s)$ and $B(s)$ as defined in (6.2).

6.1 Generalized Brownian Motions

6.1.1 Statistical Properties of Generalized GBM

To better understand the consequences resulting from the deviations of the stochastic behavior of firm's asset value $S(t)$ from the geometric Brownian motion, we consider here the stochastic process

$$S(t) = \omega[Y(t)], \tag{6.5}$$

equal to an arbitrary function $\omega(\cdot)$ of the Wiener process with drift $Y(t)$ defined by (2.6)

$$dY(t) = c\,dt + b\,dW(t), \qquad Y(0) = 0.$$

We will suppose in what follows that $s = \omega(y)$ is a smooth, positive, and increasing function of y, $y \in \mathbb{R}$. Accordingly, associated to (6.5), there exists a single-valued and increasing inverse function $y = \tilde{\omega}(s)$. We will suppose for definiteness that the following inequalities are true

$$\tilde{\omega}(s) > 0 \quad \text{if} \quad s > s_0 \quad \text{and} \quad \tilde{\omega}(s) < 0 \quad \text{for} \quad s < s_0. \tag{6.6}$$

The set of stochastic processes (6.5) includes the geometric Brownian motion as a particular case for

$$\omega(y) = s_0 e^y \qquad \Longleftrightarrow \qquad \tilde{\omega}(s) = \ln\left(\frac{s}{s_0}\right). \tag{6.7}$$

The semi-geometric Brownian motion (6.78) is also a particular case of the stochastic processes (6.5). We refer to the stochastic process (6.5) as a generalized geometric Brownian motion.

Introducing the notation

$$c(s) := \frac{c}{\tilde{\omega}'(s)}, \tag{6.8}$$

it is convenient to apply Ito's lemma to derive the stochastic differential equation that gives the dynamics of the generalized GBM (6.5)

$$dS(t) = \underbrace{\left(c[S(t)] + \frac{b^2}{4c^2}\frac{dc^2[S(t)]}{dS(t)}\right)}_{a[S(t)]} dt + b\,\frac{c[S(t)]}{c}\,dW(t) \qquad S(0) = s_0. \tag{6.9}$$

For the sequel, it is useful to determine the relation between the density $\phi(y;t)$ given by (2.8) of the Brownian motion with drift $Y(t)$ given by (2.6) and the density $f(s;t)$ of $S(t)$ defined as a function of $Y(t)$ according to (6.5). Let $F(s;t)$ and $H(y;t)$ be the cumulative distributions of the stochastic processes $S(t)$ and $Y(t)$. By definition,

$$F(s;t) := \Pr\{S(t) < s\} \qquad H(y;t) := \Pr\{Y(t) < y\}. \tag{6.10}$$

The following inequalities are equivalent:

$$S(t) < s \qquad \Longleftrightarrow \qquad Y(t) < \tilde{\omega}(s), \tag{6.11}$$

which leads to

$$F(s;t) = H(\tilde{\omega}(s);t). \tag{6.12}$$

Differentiating both sides of (6.12) with respect to s, we obtain the sought relation between the distributions $f(s;t)$ and $\phi(y;t)$:

$$f(s;t) = \tilde{\omega}'(s)\,\phi(\tilde{\omega}(s);t). \tag{6.13}$$

Substituting here $\phi(y;t)$ given by (2.8), the distribution of the stochastic process (6.5) reads

$$f(s;t) = \frac{\tilde{\omega}'(s)}{\sqrt{2\pi b^2 t}} \exp\left(-\frac{(\tilde{\omega}(s) - ct)^2}{2b^2 t}\right). \tag{6.14}$$

Replacing this distribution in the integral (3.18) yields

Proposition 6.1.1. *Under assumptions 1-3 articulated in Chap. 3. There exists a stationary mean density of firm asset values, whose behavior is described by the stochastic process (6.5). It reads*

$$g(s) = \frac{\nu}{|c|} \tilde{\omega}'(s) \exp\left(\frac{1}{b^2}\left[c\tilde{\omega}(s) - |c\tilde{\omega}(s)|\right]\right). \tag{6.15}$$

Accounting for condition (6.6), relation (6.15) can be rewritten as

$$g(s) = \frac{\nu}{|c|} \tilde{\omega}'(s) \exp\left(-\frac{|c| - c}{b^2} \int_{s_0}^{s} \tilde{\omega}'(u)\, du\right), \qquad s \geq s_0, \tag{6.16}$$

so that, by virtue of Theorem A3.3 on regularly varying functions in Embrechts et al. (1997), we can state

Proposition 6.1.2. *When $\lambda = \frac{2c}{b^2} < 0$, the stationary mean density of firms assets values, whose behavior is described by the stochastic process (6.5) follows an asymptotic power law if and only if*

$$\lim_{s \to \infty} s \cdot \tilde{\omega}'(s) = \alpha > 0. \tag{6.17}$$

In such a case, the tail index of the power law is $m = \alpha \cdot |\lambda|$

Remark 6.1.1. When λ (or c) is positive, the stationary mean density simplifies to

$$g(s) = \frac{\nu}{|c|} \tilde{\omega}'(s), \qquad s \geq s_0, \tag{6.18}$$

and it follows an asymptotic power law if $\tilde{\omega}'(s)$ is itself a power law. But, assuming for instance that $\tilde{\omega}'(s) = s^{-2}$, so that Zipf's law holds, we get $\tilde{\omega}(s) = s_0^{-1} - s^{-1}$ in order to satisfy assumption (6.6) and $\omega(y) = \left(s_0^{-1} - y\right)^{-1}$, which is neither a positive nor an increasing function on $\mathbb{R}$ as initially assumed. Consequently, Zipf's law cannot be obtained when $\lambda > 0$.

As a byproduct of Proposition 6.1.2, we get

Corollary 6.1.1. *Under assumptions 1-3 articulated in Chap. 3. The stationary mean density of firms assets values, whose behavior is described by the stochastic process (6.5) follows Zipf's law if and only if*

$$\lim_{s \to \infty} s \cdot \tilde{\omega}'(s) = -\frac{1}{\lambda} > 0. \tag{6.19}$$

Accounting for relations (6.2) and (6.9), Corollary 6.1.1 leads to

$$A(s) = \frac{c}{s \cdot \tilde{\omega}'(s)} \left(1 - \frac{1}{\lambda} \cdot \frac{\tilde{\omega}''(s)}{\tilde{\omega}'(s)^2}\right) \to 0, \qquad \text{as } s \to \infty, \tag{6.20}$$

$$B(s) = \frac{b}{s \cdot \tilde{\omega}'(s)} \to \frac{2|c|}{b}, \qquad \text{as } s \to \infty, \tag{6.21}$$

which means that

Proposition 6.1.3. *When the dynamic of firm's asset values follow generalized-GBMs, the stationary mean distribution of firm sizes follows Zipf's law if and only if both Gibrat's law and the balance condition hold asymptotically, i.e.,* $A(s) \to 0$ *and* $B(s) \to \text{constant} \in \mathbb{R}_+$ *as* $s \to \infty$.

6.1.2 Deterministic Skeleton of the Mean Density $g(s)$ Given by a Generalized-GBM

Taking into account inequalities (6.6), we observe that, in the deterministic limit ($b \to 0$), expression (6.15) reduces to

$$g(s) = \frac{\nu}{|c|} \tilde{\omega}'(s) \mathbf{1}[c(s - s_0)]. \tag{6.22}$$

In the particular case where $c > 0$, expression (6.22) becomes

$$g(s) = \frac{\nu}{c} \tilde{\omega}'(s) \, \mathbf{1}(s - s_0), \qquad c > 0, \tag{6.23}$$

which is nothing but expression (6.18) when $s \geq s_0$. This result is also reminiscent of those obtained in Sect. 4.8.

In order to reveal the geometrical meaning of relation (6.23), let us first discuss the deterministic limit of the stochastic process $S(t)$ given by (6.5). This limit amounts to replacing the Brownian motion $Y(t)$ with drift by its deterministic limit $Y_0(t) = ct$, yielding the deterministic process $S_0(t) = \omega(ct)$. The differential equation to which this process satisfies is obtained by differentiating $S_0(t) = \omega(ct)$ with respect to t to obtain

$$\frac{dS_0(t)}{dt} = c\omega'(ct) = c\omega'(Y_0(t)), \tag{6.24}$$

where $\omega'(y)$ is the derivative of the function $\omega(y)$ with respect to y. Replacing in the r.h.s. of (6.24) ct by $\tilde{\omega}[S_0(t)]$ and using the identity

$$\omega'[\tilde{\omega}(s)] \equiv \frac{1}{\tilde{\omega}'(s)}, \tag{6.25}$$

we rewrite (6.24) in the form of the differential equation

$$\frac{dS_0(t)}{dt} = c[S_0(t)], \tag{6.26}$$

where $c(s)$ has been defined by (6.8).

From the assumption that $\tilde{\omega}(s)$ is an increasing function, so that its derivative $\tilde{\omega}'(s)$ is positive, the velocity $c(s)$ given by (6.8) of the deterministic process $S_0(t)$ should be positive, if c is positive, and negative, if c is negative. Either way, we can write

$$\tilde{\omega}'(s) = \left| \frac{c}{c(s)} \right| = \frac{c}{c(s)}. \tag{6.27}$$

Then, using relation (6.27), equality (6.15) can be rewritten in the geometrically transparent form

$$g(s) = \frac{\nu}{c(s)} \, 1(s - s_0), \tag{6.28}$$

which coincides with relation (4.67).

6.1.3 *Size Dependent Drift and Volatility*

We now treat a specific example of the generalized geometric Brownian motion introduced above, inspired by the model of mutual fund sizes introduced by Schwarzkopf and Farmer (2008). Let $U(t)$ be the logarithm of a firm's asset value

$$S(t) = s_0 e^{U(t)}, \qquad U(t) = \ln\left(\frac{S(t)}{s_0}\right), \tag{6.29}$$

which is assumed to follow a stochastic process, whose drift and diffusion are functions of size according to

$$c(u) = c(\varkappa v(u) + 1), \qquad b(u) = bc(u)/c = b(\varkappa v(u) + 1). \tag{6.30}$$

Here, $v(u)$ is a decreasing function of u, such that $v(0) = 1$ and $v(\infty) = 0$. In the interpretation of Schwarzkopf and Farmer (2008), $v(u)$ describes the dependence of the money flux to mutual funds as a function of their sizes, or just the fact that it is more difficult to raise large sums of money in absolute terms. In what follows we will suppose, for concreteness, that $v(u)$ is exponential

$$v(u) = e^{-\vartheta u}, \tag{6.31}$$

so that relations (6.30) transform to the following relations, which are similar to those used by Schwarzkopf and Farmer (2008),

$$
\begin{aligned}
c(u) &= c(\varkappa e^{-\vartheta u}+1) = c_0\, s^{-\vartheta} + c_\infty, \qquad c_0 = c\varkappa s_0^{\vartheta}, \quad c_\infty = c, \\
b(u) &= b(\varkappa e^{-\vartheta u}+1) = b_0\, s^{-\vartheta} + b_\infty, \qquad b_0 = b\varkappa s_0^{\vartheta}, \quad b_\infty = b.
\end{aligned}
\tag{6.32}
$$

The data on mutual funds analyzed by Schwarzkopf and Farmer (2008) suggest the following orders of magnitudes for the parameters of the model:

$$
\vartheta \simeq 0.3, \qquad \frac{b_0}{b_\infty} = \varkappa s_0^{\vartheta} \simeq 6 \div 10 \quad \Rightarrow \quad \varkappa^{1/\vartheta} s_0 \simeq 4 \cdot 10^2 \div 2 \cdot 10^3. \tag{6.33}
$$

It follows from (6.30), (6.31), and from previous relations derived in this chapter, that $U(t)$ and $S(t)$ are generalized GBMs, such that

$$
\begin{aligned}
\tilde{\omega}(s) &= \frac{1}{\vartheta} \ln\left(\frac{1}{s_0^{\vartheta}}\, \frac{s^{\vartheta}+\varkappa s_0^{\vartheta}}{1+\varkappa}\right), \\
\tilde{\omega}'(s) &= \frac{s^{\vartheta-1}}{s^{\vartheta}+\varkappa s_0^{\vartheta}}.
\end{aligned}
\tag{6.34}
$$

Substituting the last two equalities into (6.15), we obtain the mean density of firm asset values in the case of size-dependent drift and volatility given by (6.32)

$$
g(s) = \frac{\nu}{|c|} (1+\varkappa)^{\frac{|c|-c}{\vartheta b^2}}\, s_0^{\frac{|c|-c}{b^2}}\, \frac{s^{\vartheta-1}}{\left(s^{\vartheta}+\varkappa s_0^{\vartheta}\right)^{1+\frac{|c|-c}{\vartheta b^2}}}, \qquad s > s_0. \tag{6.35}
$$

This mean density is characterized by two power asymptotics. For small asset values, where $\varkappa s_0^{\vartheta} \gg s^{\vartheta}$, one has

$$
g(s) \sim s^{\vartheta-1}, \qquad s_0 < s \ll \varkappa^{1/\vartheta} s_0. \tag{6.36}
$$

For large asset values, $g(s) \sim s^{-\left(\frac{|c|-c}{b^2}+1\right)}$, so that Zipf's law is recovered if the balance condition $a = 0$, i.e., $c = -\frac{b^2}{2}$, is satisfied. Figure 6.1 shows the mean density of firm sizes given by (6.35) when the latter condition holds.

6.2 Diffusion Process with Constant Volatility

In our previous explanation of the statistical properties of the GBM (2.10), expression (2.11) was essentially based on the statistical properties of the Wiener process $Y(t, c, b)$ with drift defined by (2.6), which satisfies the stochastic equation

$$
dY(t, c, b) = c\, dt + b\, dW(t), \qquad Y(0) = 0. \tag{6.37}
$$

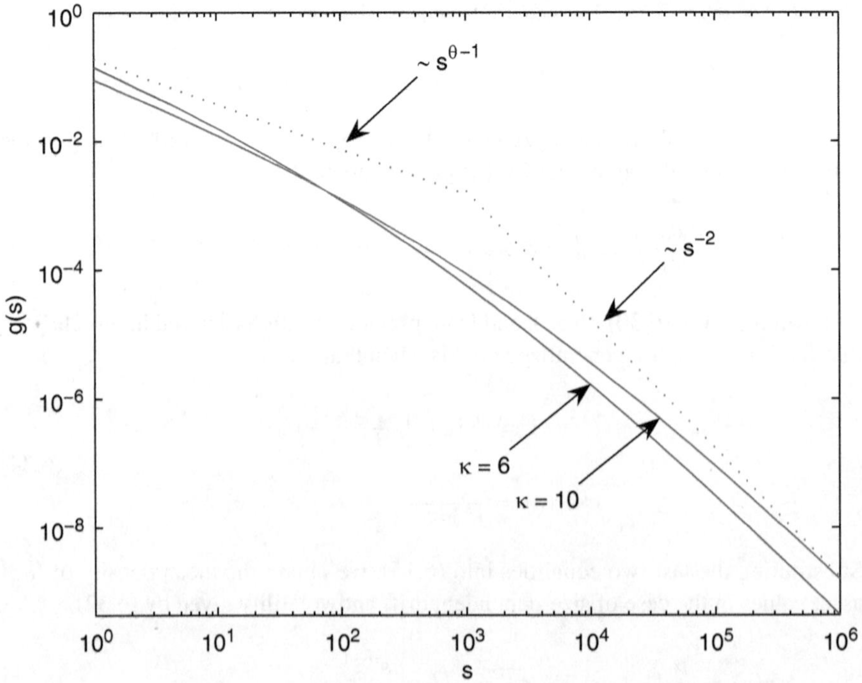

Fig. 6.1 Log-log plots of the mean density (6.35) for $s_0 = 1$, $\vartheta = 0.3$ and $\varkappa = 6, 10$. One can observe clearly the two power asymptotics (6.36) and Zipf's law $g(s) \sim s^{-2}$

The distinctive peculiarity of the process defined by (6.37) is the constancy of the volatility coefficient $b = \text{const}$. It will be clear later on, that the analogous diffusion process $Z(t)$ with constant (unit) volatility which satisfies the stochastic equation

$$dZ(t) = c[Z(t)]\,dt + dW(t), \qquad Z(0) = z_0, \tag{6.38}$$

plays an essential role in discussing the statistical properties of process $S(t)$ which satisfies the general stochastic equation (6.1). Bearing this in mind, let us find a relation expressing the solution to (6.1) via the solution to the auxiliary stochastic equation (6.38). To find the mentioned relation, we first rewrite the diffusion equation (2.39) in a form more convenient for the following analysis

$$\frac{\partial f(s;t)}{\partial t} + \frac{\partial[\tilde{c}(s)f(s;t)]}{\partial s} = \frac{1}{2}\frac{\partial}{\partial s}\left(b(s)\frac{\partial[b(s)f(s;t)]}{\partial s}\right), \tag{6.39}$$

where

$$\tilde{c}(s) := a(s) - \frac{1}{2}b(s)\frac{db(s)}{ds}. \tag{6.40}$$

We will suppose in what follows that $b(s)$ is positive everywhere and is a differentiable function of the argument s.

We introduce the change of variable

$$z = z(s) := \int_{s_1}^{s} \frac{du}{b(u)} \qquad \Rightarrow \qquad \frac{ds(z)}{dz} = b(s(z)). \tag{6.41}$$

Multiplying all terms of (6.39) by $b(s)$ and taking into account the following operational relation

$$b(s)\frac{\partial}{\partial s} \qquad \Longleftrightarrow \qquad \frac{\partial}{\partial z}, \tag{6.42}$$

we obtain the diffusion equation

$$\frac{\partial h(z;t)}{\partial t} + \frac{\partial [c(z)h(z;t)]}{\partial z} = \frac{1}{2}\frac{\partial^2 h(z;t)}{\partial z^2}, \tag{6.43}$$

with

$$h(z;t) = b(s)f(s;t)\Big|_{s=s(z)}, \tag{6.44}$$

where $s(z)$ is the inverse of function $z(s)$ defined in (6.41), and

$$c(z) := \frac{\tilde{c}(s)}{b(s)}\Big|_{s=s(z)} = \left(\frac{a(s)}{b(s)} - \frac{1}{2}\frac{db(s)}{ds}\right)\Big|_{s=s(z)}. \tag{6.45}$$

If the diffusion equation (2.39) is supplemented by the initial condition (2.40), then one has to supplement (6.43) by the analogous condition

$$h(z;0) = \delta(z - z_0), \qquad \text{where} \qquad z_0 = z(s_0). \tag{6.46}$$

In turn, if the solution to the diffusion equation (2.39) satisfies the condition of firm's exit at level s_1

$$f(s_1;t) = 0, \tag{6.47}$$

then it follows from (6.44) and (6.41) that the solution to (6.43) should satisfy the similar condition

$$h(0;t) = 0. \tag{6.48}$$

Reversely, if one knows the solution to the initial-boundary problem (6.43), (6.46), (6.48), then one can obtain the solution to the initial-boundary problem (2.39), (2.40), (6.47), using the inverse of relation (6.44):

$$f(s;t) = \frac{h[z(s);t]}{b(s)}. \tag{6.49}$$

As the pdf $h(z;t)$ of the solution to the stochastic equation (6.38) satisfies the diffusion equation (6.43), one may interpret the above relations (6.44) and (6.49) in the sense that the diffusion processes $S(t)$ and $Z(t)$ are tied by the equivalence relation

$$S(t) \equiv s[Z(t)] \qquad \Longleftrightarrow \qquad Z(t) \equiv z[S(t)]. \tag{6.50}$$

To sum up, we state

Proposition 6.2.1. *The solution to the stochastic differential equation* (6.1) *supplemented by the exit condition at the level* s_1 *is* $S(t) = s(Z(t))$*, where* $Z(t)$ *is the stochastic process solution to the stochastic differential equation* (6.38) *supplemented by the exit condition at the level* $z = 0$ *and* $s(\cdot) := z^{-1}(\cdot)$ *is defined by* (6.41).

6.3 Steady-State Density of Firm's Asset Values in the Presence of Deviations from Gibrat's Law

Let us discuss the influence of deviations from Gibrat's law on the steady-state mean density $g(s)$ given by (3.18), supposing that the firm's asset value $S(t)$ satisfies the general stochastic equation (6.1). The corresponding pdf $f(s;t)$ is expressed via the solution to the diffusion equation (6.43) by equality (6.49). Integrating both sides of this equality with respect to t over the interval $t \in [0, \infty)$, we obtain, provided that $\int_0^\infty f(s,t)dt < \infty$, for all s

$$g(s) = \frac{\nu}{b(s)}\, \eta[z(s)], \tag{6.51}$$

where

$$\eta(z) := \int_0^\infty h(z;t)dt \tag{6.52}$$

is, analogously to (4.5), the distribution of mean time interval durations for which the process $Z(t)$ with unit volatility satisfying the stochastic equation (6.38) is above some arbitrary level z.

Integrating the diffusion equation (6.43) with respect to t over the interval $t \in [0, \infty)$, and assuming that due to diffusion, drift and/or death of the diffusion process $Z(t)$ upon first touching the zero level, the pdf $h(z;t)$ satisfies the limiting condition

$$\lim_{t\to\infty} h(z;t) = 0, \tag{6.53}$$

we obtain the following equation for $\eta(z)$

$$\frac{1}{2}\frac{d^2\eta(z)}{dz^2} - \frac{d[c(z)\eta(z)]}{dz} = -\delta(z - z_0). \tag{6.54}$$

Dirac's delta function in the r.h.s. of the last equation takes into account the initial condition (6.46). In addition, to take into account the killing of the process $Z(t)$ at zero level, we have to supplement (6.46) by the boundary condition

$$\eta(z=0)=0, \tag{6.55}$$

following from condition (6.48).

It is easy to show that the general solution to (6.54), together with boundary condition (6.55), is equal to

$$\eta(z)=2\begin{cases}(1-D)\mathcal{S}(z,0), & 0<z<z_0,\\ \mathcal{S}(z_0,0)\;\mathcal{Q}(z,z_0)-D[\mathcal{S}(z_0,0)\mathcal{Q}(z,z_0)+\mathcal{S}(z,z_0)], & z>z_0,\end{cases} \tag{6.56}$$

where

$$\mathcal{Q}(x,y):=\exp\left(2\int_y^x c(z)dz\right),\qquad \mathcal{S}(x,y):=\int_y^x \mathcal{Q}(x,z)dz. \tag{6.57}$$

Solution (6.56) depends on an arbitrary constant D. Its arbitrariness is the consequence of a lack of boundary condition prescribing the asymptotic behavior of the distribution $\eta(z)$ as $z\to\infty$.[1] Before seeking the value of the constant D, note that the asymptotic behavior of $\eta(z)$ as $z\to\infty$ strongly depends on the behavior of the realizations of the diffusion process $Z(t)$ under investigation. In the following section, we will discuss the behavior of the diffusion process $Z(t)$ in detail, while in this section we will restrict ourselves to discussing the financial meaning of the constant D appearing in the r.h.s. of expression (6.56).

One may interpret the diffusion equation (6.43) as a continuity equation and rewrite it in the form

$$\frac{\partial h(z;t)}{\partial t}+\frac{\partial \mathcal{F}(z;t)}{\partial z}=0, \tag{6.58}$$

where

$$\mathcal{F}(z;t):=c(z)h(z;t)-\frac{1}{2}\frac{\partial h(z;t)}{\partial z}. \tag{6.59}$$

The function $\mathcal{F}(z;t)$ has a transparent geometrical interpretation. It is the mean flow of the realizations of the diffusion process $Z(t)$ through the level z at current time t. It is easy to check that the discussed arbitrary constant D is nothing but the mean flow above the initial level $z>z_0$ integrated over all times

$$D=\int_0^\infty \mathcal{F}(z;t)dt,\qquad z>z_0. \tag{6.60}$$

Proposition 6.3.1. *Under the assumption that the integrated flow is equal to zero for $s>s_0$, then $D=0$ and the steady-state mean density reads*

[1] The fact that $z\to+\infty$ as $s\to+\infty$ is ensured by assumption (6.3) and definition (6.41).

$$g(s) = \frac{2\nu}{b(s)} \times \begin{cases} \mathcal{S}(z(s),0), & s_1 < s < s_0, \\ \mathcal{S}(z(s_0),0)\ \mathcal{Q}(z(s),z(s_0)), & s > s_0. \end{cases} \tag{6.61}$$

Below, we will discuss why and under which conditions the equality $D = 0$ may be true.

6.4 Integrated Flow

As we just showed, the solution (6.56) to the boundary condition problem (6.54), (6.55) contains the constant D, which is equal to the mean integrated flow (6.60) above some initial level $z > z_0$. In order to determine the value of D for some cases that are important for our purposes, we will use the following conjecture:

Conjecture 6.4.1. If the realizations of the diffusion process $Z(t)$, satisfying the stochastic equation (6.38), are bounded from above, then the integrated flow (6.60) at $z > z_0$ is almost surely equal to zero ($D = 0$).

We interpret the conjectured boundedness of the realizations of the diffusion process $Z(t)$ in the sense that the cdf

$$F_+(z) := \Pr\{Z_+ < z\} \tag{6.62}$$

of the absolute maximal values of the realizations

$$Z_+ := \max_{t\in[0,\infty)} Z(t) \tag{6.63}$$

tends to one as z tends to infinity:

$$\lim_{z\to\infty} F_+(z) = 1. \tag{6.64}$$

The following calculations support the above conjecture. Consider the distribution $\eta(z)$ of mean life durations in the simplest case of constant drift $c(z) = c =$ const. We can obtain $\eta(z)$ by direct calculation of the integral (6.52). Firstly, we need to find the solution to the diffusion equation

$$\frac{\partial h(z;t)}{\partial t} + c\frac{\partial h(z;t)}{\partial z} = \frac{1}{2}\frac{\partial^2 h(z;t)}{\partial z^2} \tag{6.65}$$

with initial condition (6.46) and boundary condition (6.48). Using the reflection method (Morse and Feshbach, 1953), we obtain the expression [analogously to (5.13)]

$$h(z;t) = \varphi(z - z_0;t) - e^{-2cz_0}\varphi(z + z_0;t), \tag{6.66}$$

where

$$\varphi(z;t) = \frac{1}{\sqrt{2\pi t}} \exp\left(-\frac{(z-ct)^2}{2t}\right). \tag{6.67}$$

Substituting (6.66) into relation (6.52) and using the integral identity (3.23), we obtain

$$\eta(z) = \frac{1}{|c|} e^{c(z-z_0)} \left[e^{-|c(z-z_0)|} - e^{|c|(z+z_0)}\right]. \tag{6.68}$$

Thus, for the case $c < 0$, we have

$$\eta(z) = \frac{1}{|c|} \begin{cases} 1 - e^{2cz}, & 0 < z < z_0, \\ e^{2cz}\left(e^{-2cz_0} - 1\right), & z > z_0, \end{cases} \qquad c < 0, \tag{6.69}$$

while for $c > 0$ we obtain

$$\eta(z) = \frac{1}{c} \begin{cases} e^{-2cz_0}\left(e^{2cz} - 1\right), & 0 < z < z_0, \\ 1 - e^{-2cz_0}, & z > z_0, \end{cases} \qquad c > 0. \tag{6.70}$$

We note that expression (6.69) does obey the above conjecture. Indeed, the diffusion process $Z_1(t)$, whose pdf obeys the initial-value problem (6.65), (6.46), (6.48), satisfies the stochastic equation

$$dZ_1 = c\,dt + dW(t), \qquad Z_1(0) = z_0. \tag{6.71}$$

The corresponding cdf (6.62) of the maximal value of the diffusion process $Z_1(t)$ is equal to

$$F_+(z) = 1 - \int_0^\infty h(z;t)dt, \tag{6.72}$$

where $h(z;t)$ is described by expression (6.65). Calculating the integral, we obtain

$$F_+(z) = 1 - e^{2c(z-z_0)}, \qquad z > z_0, \qquad c < 0. \tag{6.73}$$

Obviously this cdf tends to one as $z \to \infty$. Thus, the diffusion process $Z_1(t)$ is almost surely bounded from above. This means, according to our conjecture, that the integrated flow for $z > z_0$ should be equal to zero ($D = 0$). In fact, it is easy to show that the distribution $\eta(z)$ given by (6.69) of mean time durations coincides with expression (6.61) for $c =$ const., corresponding to zero integrated flow $D = 0$.

Based on our conjecture, one may use expression (6.61) for the distribution of mean time duration $\eta(z)$ if the realizations of the solutions to the stochastic equation (6.38) are almost surely bounded from above. Some necessary conditions for realizations of the diffusion process $Z(t)$ to be bounded from above are detailed below.

Proposition 6.4.1. *Suppose that there exists a negative constant $c < 0$ such that the following inequality is valid*

$$c(z) < c < 0, \qquad 0 < z < \infty, \tag{6.74}$$

then, almost surely, the stochastic process $S(t)$ given by (6.50) is bounded from above.

To prove this proposition, we first note that, if inequality (6.74) is true, then, for the same initial conditions $Z(0) = Z_1(0) = z_0$ and for the same realization of the Wiener process $W(t)$, the corresponding solutions to the stochastic equations (6.71), (6.38) obey the inequality

$$Z(t) < Z_1(t), \qquad t > 0. \tag{6.75}$$

In fact, it follows from (6.71), (6.38) that the residual

$$V(t) := Z(t) - Z_1(t) \tag{6.76}$$

satisfies the inequality

$$dV(t) = c[Z(t)] - c < 0 \qquad \Rightarrow \qquad V(t) < 0, \quad t > 0. \tag{6.77}$$

This means in turn that inequality (6.75) is valid.

We already showed that, if $c < 0$ then, the diffusion process $Z_1(t)$ is almost surely bounded from above. Thus, it follows from inequality (6.75) that $S(t)$ given by (6.50) is also bounded from above. □

There is a non-rigorous but geometrically transparent explanation of why, for the above-bounded diffusion process $Z(t)$, the integrated flow at $Z > z_0$ is equal to zero. Roughly speaking, this kind of process $Z(t)$ crosses any level $z > z_0$ an even number of times, so that the mean flows up and down occurring over an infinite time interval $t \in [0, \infty)$ compensate each other exactly. In other words, the equality $D = 0$ at $z > z_0$ expresses an integrated balance condition for the diffusion process $Z(t)$ bounded from above. Figure 6.2 presents schematic pictures illustrating why the integrated flow D is equal to zero for any strictly negative $c(z)$ and for $z > z_0$.

6.5 The Semi-Geometric Brownian Motion

Using the approach developed above, based on relations between the statistical properties of the diffusion process $S(t)$ and the auxiliary process $Z(t)$ with constant (unit) volatility, let us consider the statistical properties of the simplest diffusion process that demonstrates a deviation from Gibrat's law:

Definition 6.5.1. The diffusion process $S(t)$ with drift and diffusion coefficients:

$$a(s) = a \cdot (s + \kappa), \qquad b(s) = b \cdot (s + \kappa), \qquad \kappa > 0, \tag{6.78}$$

is called a *semi-geometric Brownian motion*.

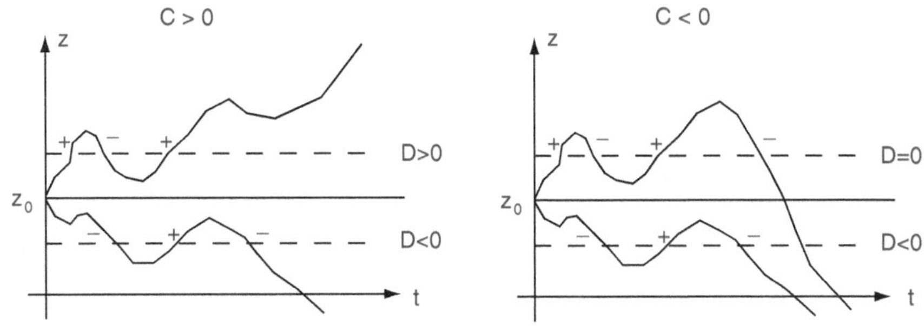

Fig. 6.2 A schematic illustration of the value of the integrated flow D (6.60) for cases of positive and negative regular drift $c(z)$. In the first case (*left picture*), the regular upward drift prevails over the diffusion part, so that a nonzero fraction of realizations of the process $Z(t)$ are not killed at the zero level and escape. These realizations tend to infinity with the course of time. Such realizations cross any level $z > z_0$ an odd number of times, with the number of upward crossings being one unit more than the number of downward crossings. As a result, the integrated flow D defined by (6.60) becomes positive. In contrast, all realizations, which are killed at the zero level, cross any level $z < z_0$ downwards one more time than they cross the same level while going upwards. As a result, the integrated flow under the level z_0 is negative. For the case of strictly negative $c(z)$ (*right picture*), the number of crossings of a level $z > z_0$ downwards and upwards are the same, so that the integrated flow D turns out to be exactly zero

Obviously, it reduces to the regular GBM for the particular case $\kappa = 0$.

The main distinction between the semi-GBM and the regular GBM is the fact that both the drift $a(s)$ and the volatility $b(s)$ do not vanish as $s \to 0$. This means in particular that, unlike for the regular GBM, there is a nonzero probability of crossing the zero level ($s = 0$), due to a positive zero-level diffusion coefficient

$$b_0 := b(0) = b\kappa. \tag{6.79}$$

In contrast, for $\kappa = 0$, the diffusion coefficient vanishes at $s = 0$ preventing crossing by so-to-speak "freezing" the diffusion dynamics for small s values. Figure 6.3 plots the ratio $B(s)$ defined by (6.2) for the case (6.78), demonstrating the essential deviation of the semi-GBM from Gibrat's law for $s \lesssim \kappa$.

Substituting the diffusion coefficient $b(s)$ given in (6.78) into relation (6.41), we obtain a function that maps the semi-GBM into a diffusion process with unit volatility (and vice versa)

$$z(s) = \frac{1}{b} \ln\left(\frac{s+\kappa}{s_1+\kappa}\right) \qquad \Longleftrightarrow \qquad s(z) = (s_1+\kappa)e^{bz} - \kappa. \tag{6.80}$$

It is seen from (6.45) that the auxiliary process with constant volatility is nothing but a diffusion process $Z_1(t)$ satisfying the stochastic equation (6.71), where

$$c = \frac{a}{b} - \frac{b}{2}. \tag{6.81}$$

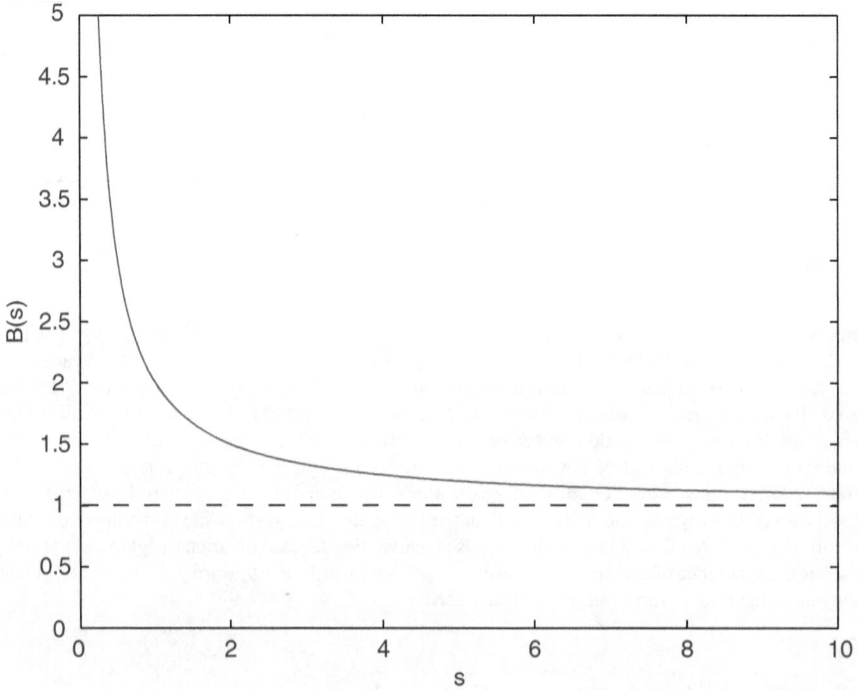

Fig. 6.3 A plot of the ratio $B(s)$ defined by (6.2), for $b = 1$ and $\kappa = 1$, demonstrating the deviation of the semi-GBM from the regular GBM

In turn, $Z_1(t)$ is tied to a Wiener process with drift defined by (2.6) by the obvious relation

$$Z_1(t) = z_0 + Y(t, c, 1), \qquad z_0 = \frac{1}{b} \ln\left(\frac{s_0 + \kappa}{s_1 + \kappa}\right). \tag{6.82}$$

Thus, it follows from (6.80), (6.82) and (6.50), that there is a linear relation between the semi-GBM and the regular GBM $X(t, c, 1)$ defined by (2.11):

$$S(t) = (s_0 + \kappa)X(t, c, 1) - \kappa. \tag{6.83}$$

Substituting the auxiliary functions $\eta(z)$ defined by (6.69) and $z(s)$ defined by (6.80) into (6.51), and taking into account relation (6.81), we obtain the steady-state mean density of firm's asset values for the case where their stochastic behavior is described by the semi-GBM.

Proposition 6.5.1. *The steady-state mean density of firm's size when the stochastic behavior of firm's size is described by the semi-GBM with $\lambda < 0$ is*

$$g(s) = \frac{\mathcal{N}}{(s+\kappa)} \begin{cases} 1 - \left(\frac{s+\kappa}{s_1+\kappa}\right)^{\lambda}, & s_1 < s < s_0, \\ \left(\frac{s+\kappa}{s_0+\kappa}\right)^{\lambda} \left[1 - \left(\frac{s_0+\kappa}{s_1+\kappa}\right)^{\lambda}\right], & s > s_0, \end{cases} \tag{6.84}$$

where

$$\mathcal{N} = \frac{2\nu}{b^2|\lambda|}. \tag{6.85}$$

Note in conclusion that, for the case $\kappa = 0$, expression (6.84) reduces to the already known Proposition 5.7.1.

Proposition 6.5.1 describes the influence on the steady-state density $g(s)$ of firm's sizes of two distinct deviations from Gibrat's law, associated with the two scales, s_1 and κ, which are present in the relation. The former scale s_1 takes into account the violation of Gibrat's law due to a non-zero asset value at which firms exit. The latter scale κ describes the deviations to Gibrat's law of the diffusion process $S(t)$ for small sizes, before firms disappear.

Taking $s_1 = 0$ in (6.84) allows us to isolate the influence of κ alone. In particular, for the balanced case $\lambda = -1$, for which Gibrat's law leads to Zipf's law, we have

$$g(s) = \frac{2\nu}{b^2|\lambda|(s+\kappa)^2} \begin{cases} s, & 0 < s < s_0, \\ s_0, & s > s_0. \end{cases} \tag{6.86}$$

Let us stress that the steady-state mean density $g(s)$ given by (6.86) depends on two scale parameters, essentially influencing the shape of the steady-state mean density. The first one is κ, responsible for the non-Gibrat's behavior of realizations of the diffusion process $S(t)$. The second one is the initial asset value s_0 of all firms.

Assuming that all firms were born with the same initial asset value is of course too restrictive. To weaken this assumption, suppose that the initial asset value s_0 is a random variable whose complementary cdf is equal to $\bar{F}_{s_0}(s)$. Averaging $g(s)$ given by (6.86) over the statistic of the random initial asset value, we obtain the new density of firm's sizes, which takes into account the distribution of their sizes at the times of their creation:

$$g(s) = \frac{2\nu}{b^2|\lambda|(s+\kappa)^2} \int_0^s \bar{F}_{s_0}(u)du. \tag{6.87}$$

As a plausible distribution of the random asset value s_0 at birth, we consider the gamma distribution whose complementary cumulative distribution reads

$$\bar{F}_{s_0}(s) := \bar{F}_{s_0}(s; \bar{s}, q) = \frac{1}{\Gamma(q)} \Gamma\left(q, q\frac{s}{\bar{s}}\right). \tag{6.88}$$

The parameter $\bar{s}$ here is the average initial asset value $\bar{s} = \mathrm{E}[s_0]$. The other parameter q controls the width of the gamma distribution. In particular, for $q \to \infty$, the random initial asset value s_0 converges in probability to $\bar{s}$, which, in this sense, plays the role of a generalized initial asset value. We then have

$$\int_0^s \bar{F}_{s_0}(u; \bar{s}, q)du = \bar{s} + \frac{s}{\Gamma(q)} \Gamma\left(q, q\frac{s}{\bar{s}}\right) - \frac{\bar{s}}{\Gamma(q+1)} \Gamma\left(q+1, q\frac{s}{\bar{s}}\right), \tag{6.89}$$

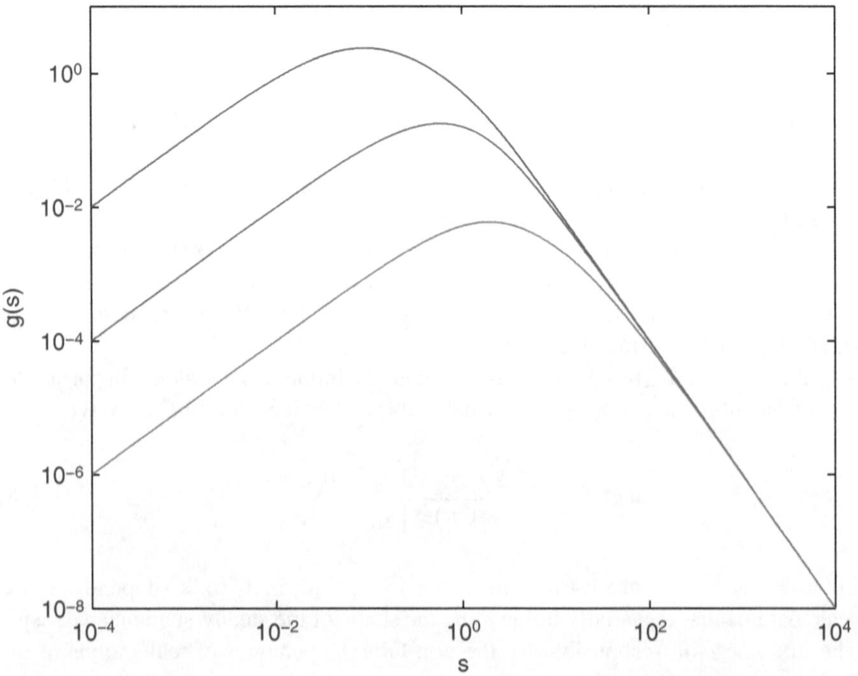

Fig. 6.4 Plots of the steady-state density $g(s, \bar{s}, \kappa, q)$ for $q = 1$, demonstrating the influence of the ratio of the two scales, κ and $\bar{s}$. The plots are drawn for $\mathcal{N} = 1$, $q = 1$ and $\bar{s} = 1$. *Top to bottom:* $\kappa = 0.1, 1, 10$

or, in terms of the complementary cdf $\bar{F}_{s_0}(s; \bar{s}, q)$ given by (6.88):

$$\int_0^s \bar{F}_{s_0}(u; \bar{s}, q)du = \bar{s} + s\,\bar{F}_{s_0}(s; \bar{s}, q) - \bar{s}\,\bar{F}_{s_0}(qs; (q+1)\bar{s}, q). \tag{6.90}$$

Let us denote by $g(s, \bar{s}, \kappa, q)$ the steady-state mean density (6.87), for the case where $\bar{F}_{s_0}(s)$ corresponds to the gamma distribution (6.88), making explicit the dependence of g on the two scales $\bar{s}$ and κ. Figure 6.4 shows $g(s, \bar{s}, \kappa, q)$ for $q = 1$ and different ratios of the scales κ and $\bar{s}$.

Figure 6.5 shows the steady-state density $g(s, \bar{s}, \kappa, q)$ for $\bar{s} = \kappa = 1$, illustrating its dependence on the parameter q, controlling the width of the gamma distribution.

6.6 Zipf's Laws When Gibrat's Law Does Not Hold

As we have demonstrated at length, when the behavior of firm's asset values is described by the GBM, then the steady-state mean density $g(s)$ obeys Zipf's law only if the balance condition $\lambda = -1$ holds true. In terms of the diffusion process $S(t)$, this means that any regular drift is absent: $a = 0$.

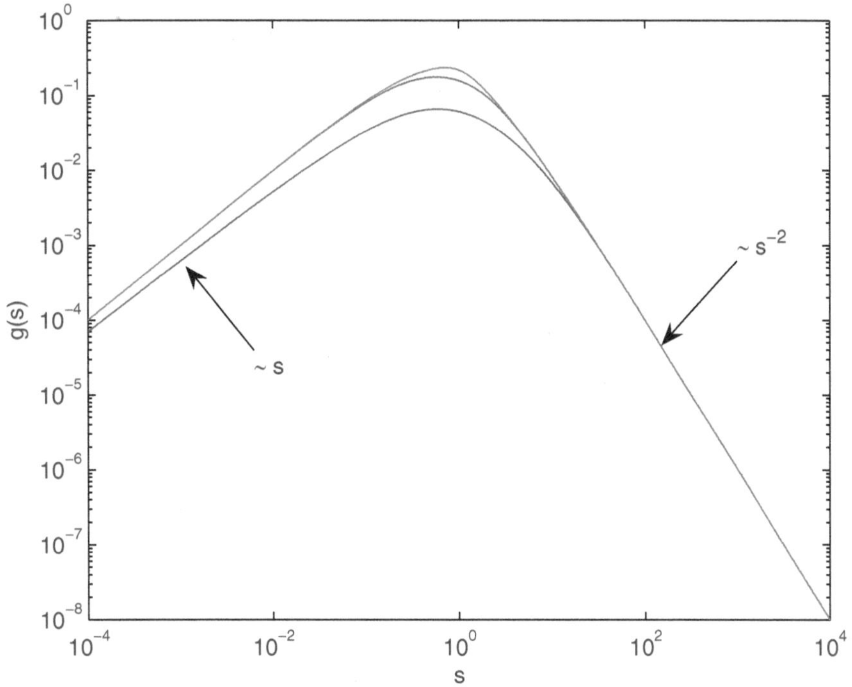

Fig. 6.5 Plots of the steady-state density $g(s, \bar{s}, \kappa, q)$ for $\mathcal{N} = 1$, $\bar{s} = \kappa = 1$. *From bottom to top:* $q = 0.1, 1, 10$

If the stochastic behavior of a firm's asset value deviates from Gibrat's law such that the asset value $S(t)$ is described by the general stochastic equation (6.1), where $a(s)$ and $b(s)$ are nonlinear functions of s, it is important to obtain the general conditions for the validity of Zipf's law, which should be imposed on the drift $a(s)$ and the diffusion coefficient $b(s)$. Since one can be interested in situations where Zipf's law holds only in an asymptotic sense, i.e., for large firms, these conditions may also be sufficient in an asymptotic sense for large sizes s. We derive below these general conditions in the case where the inequalities (6.74) holds true. Let us rewrite the first inequality in (6.74) in the form

$$c(z) < 0 \qquad \Rightarrow \qquad a(s) < \frac{1}{4}\frac{db^2(s)}{ds} \qquad (b(s) > 0). \tag{6.91}$$

If this condition is valid, then the steady-state density $g(s)$ is described by expressions (6.51) and (6.61). Eliminating the auxiliary variable z, we rewrite the steady-state density in a form more convenient for our analysis,

$$g(s) = \frac{2\nu}{b^2(s)}\,\mathcal{Q}(s)\begin{cases}\mathcal{S}(s), & 0 < s < s_0,\\ \mathcal{S}(s_0), & s > s_0,\end{cases} \tag{6.92}$$

where we have used the notations

$$\mathcal{Q}(s) := \exp\left(\int_0^s \delta(u)du\right), \qquad \mathcal{S}(s) := \int_0^s \frac{du}{\mathcal{Q}(u)}, \tag{6.93}$$

and

$$\delta(s) := \frac{2a(s)}{b^2(s)}. \tag{6.94}$$

In addition, we supposed for simplicity (but without loss of generality) in (6.92) that firms are killed at the zero level ($s_1 = 0$). Notice that condition (6.74) can be rewritten as

$$\delta(s) < \frac{d \ln b(s)}{ds} \quad \Longleftrightarrow \quad \mathcal{Q}(s) < \frac{b(s)}{b(0)}. \tag{6.95}$$

For large size s, the steady-state mean density $g(s)$ behaves like $\frac{\mathcal{Q}(s)}{b^2(s)}$ so that four cases are of interest.

Case 1

Let us first assume that

$$\exists \lim_{s\to\infty} s \cdot \delta(s) = \gamma \in \mathbb{R}. \tag{6.96}$$

According to Theorem A3.3 in Embrechts et al. (1997), $\mathcal{Q}(s)$ is a regularly varying function at infinity. This means that $\mathcal{Q}(s)$ behaves like an asymptotic power law with tail index γ. When $\gamma = 0$, one gets a slowly varying function, like a constant or a logarithm. In order for condition (6.74) – or alternatively (6.95) – to be satisfied, we need that, for large s, $b(s)$ grows at least as fast as s^γ when $\gamma > 0$ and decreases not faster than s^γ when $\gamma < 0$. If such a requirement holds, we get

$$g(s) \sim \frac{s^\gamma}{b^2(s)}, \qquad \text{as } s \to \infty. \tag{6.97}$$

We can then obtain Zipf's law if, and only if, $b(s) \sim s^{\frac{\gamma}{2}+1}$ for large s, with $\gamma \le 2$ so that $b(s)$ can grow faster than s^γ when $s \to \infty$. From (6.96), we conclude that, necessarily, $a(s) \sim \text{sgn}(\gamma) \cdot s^{\gamma+1}$, for large s.

Case 2

Let us now focus on the particular case where $\gamma = 0$ and let us in addition assume that the function $\delta(s)$ is integrable, i.e., the integral

$$\ell := \int_0^\infty \delta(u)du \tag{6.98}$$

is finite ($|\ell| < \infty$). It is interesting to remark that the asymptotic behavior (for $s \to \infty$) of the steady-state mean density $g(s)$ given by (6.92) then does not depend on the drift $a(s)$.

Indeed, in this case, the asymptotic behavior of the steady-state mean density $g(s)$ given by (6.92) is described by the relation

$$g(s) \sim \frac{2\nu e^{\ell} S(s_0)}{b^2(s)}, \qquad s \to \infty. \tag{6.99}$$

In particular, if $S(t)$ satisfies the *asymptotic diffusive Gibrat's law*, i.e., if the volatility coefficient is asymptotically linear

$$b(s) \simeq b \cdot s, \qquad s \to \infty, \tag{6.100}$$

then we recover from (6.99) an asymptotic Zipf's law for the steady-state density of firm's sizes: $g(s) \sim s^{-2}$.

In the case where the asymptotic diffusive Gibrat's law is valid, then the condition of absolute integrability of $\delta(s)$ reduces to the condition that the drift grows sufficiently slowly with s, that is, $a(s)$ is of the order of s^{α}, or in mathematical notations,

$$a(s) = \mathcal{O}\{s^{\alpha}\}, \qquad \alpha < 1, \qquad s \to \infty. \tag{6.101}$$

Case 3

We now assume that

$$\lim_{s \to \infty} s \cdot \delta(s) = +\infty. \tag{6.102}$$

Thus $\mathcal{Q}(s)$ is increasing as s goes to infinity and, according to Theorem A3.3 in Embrechts et al. (1997), it is a rapidly varying function at infinity. It means that $\mathcal{Q}(s)$ grows at infinity faster than any power law.

By virtue of condition (6.95), $b(s)$ must grow at least as fast as $\mathcal{Q}(s)$ so that $b(s)$ must also be a rapidly increasing function. Putting (6.92) and (6.95) together, we conclude that

$$g(s) < \frac{2\nu S(s_0)}{b(0)} \cdot \frac{1}{b(s)}, \tag{6.103}$$

which means that the stationary mean density $g(s)$ is a rapidly decreasing function, i.e., goes to zero faster than any power law. This situation is therefore not compatible with Zipf's law.

Case 4

We finally assume that

$$\lim_{s \to \infty} s \cdot \delta(s) = -\infty. \tag{6.104}$$

Necessarily, $\lim_{s \to \infty} a(s) < 0$ and $\mathcal{Q}(s)$ is a rapidly decreasing function. In order for the steady-state mean density of firm sizes to follow an asymptotic power law $g(s) \sim s^{-m-1}$, we need that, for s large enough

$$b(s)^2 = K s^{-m-1} \mathcal{Q}(s) + o(s), \tag{6.105}$$

so that, taking the logarithm of both sides, differentiating and multiplying by s yields

$$2\frac{s \cdot b'(s)}{b(s)} = -(m+1) + s \cdot \delta(s) + o'(s) \tag{6.106}$$

Taking the limit $s \to \infty$ we conclude, using assumption (6.104), that

$$\lim_{s\to\infty} 2\frac{s \cdot b'(s)}{b(s)} = -\infty. \tag{6.107}$$

This would mean that $\lim_{s\to\infty} b'(s) < 0$, which is impossible since $b(s) > 0$ for all s, so that $\lim_{s\to\infty} b(s) \geq 0$. Consequently, $g(s)$ cannot be an asymptotic power law under assumption (6.104).

To sum up the results of this chapter, we can state

Proposition 6.6.1. *When the dynamic of firm's asset value follows the diffusion equation* (6.1)*, provided that condition* (6.74) *holds, the mean steady-state distribution of firm sizes follows Zipf's law if and only if:*

1. $\exists \gamma \in (-\infty, 2]$, *such that* $\lim_{s\to\infty} s \cdot \frac{2a(s)}{b(s)^2} = \gamma$
2. *The volatility* $b(s)$ *is a regularly varying function at infinity with tail index* $\frac{\gamma}{2} + 1$

Remark 6.6.1. As a consequence of this proposition, we can notice that the drift $a(s)$ is also necessarily regularly varying at infinity. Indeed, by *1.*, when $\gamma \neq 0$, we get

$$a(s) \sim \frac{\gamma}{2} \cdot \frac{b^2(s)}{s}, \qquad \text{as } s \to \infty, \tag{6.108}$$

which leads to

$$\frac{a(s)}{b(s)} \sim \frac{\gamma}{2} \cdot s^{\frac{\gamma}{2}}, \qquad \text{as } s \to \infty. \tag{6.109}$$

This means that, when $\gamma < 0$, the distribution of firm sizes described by Zipf's law is more the consequence of random growth than systematic returns. In particular for large firms, volatility dominates over the instantaneous growth rate. On the contrary, when $\gamma \in (0, 2]$, Zipf's law is more the result of systematic returns than random growth.

To complete the analysis of the possible validity of Zipf's law for non-Gibrat behavior of firm's asset values, let us consider the consequence of the following power law asymptotics of the drift and volatility coefficients:

$$a(s) \simeq a \cdot s^{\alpha}, \qquad b(s) \simeq b \cdot s^{\beta}, \qquad s \to \infty. \tag{6.110}$$

In this case, the condition (6.91) of applicability of the steady-state density expression (6.92) reduces to

$$\delta < \beta s^{-\gamma-1}, \qquad \gamma := \alpha - 2\beta, \qquad \delta := \frac{2a}{b^2}. \tag{6.111}$$

This condition is qualitatively different for $\gamma < -1$, $\gamma = 1$ and $\gamma > 1$. Correspondingly, there are three qualitatively different asymptotics for the steady-state density $g(s)$ (6.92):

1. $\gamma < -1$ is a particular case of the absolute integrability condition (6.98). For $\gamma < -1$, the asymptotic relation (6.99) is true for any δ. In view of the asymptotics (6.110), we obtain $g(s) \sim s^{-2\beta}$ for $s \to \infty$.
2. For $\gamma = -1$, condition (6.111) transforms into inequality $\delta < \beta$. If this inequality holds true, then we have the asymptotics

$$Q(s) \sim s^{\delta} \qquad \Rightarrow \qquad g(s) \sim s^{\delta - 2\beta}, \qquad s \to \infty. \tag{6.112}$$

   Then, if

$$\delta - 2\beta = -2 \quad \Rightarrow \quad \alpha = \delta + 1, \quad \beta = \frac{\delta + 2}{2}, \quad \delta < 1, \tag{6.113}$$

   we obtain Zipf's law $\theta(s) \sim s^{-2}$. Note also that $\delta = 0$ corresponds to Zipf's law derived from Gibrat's law, as discussed above with respect to expression (3.27).
3. Finally, if $\gamma > -1$, then condition (6.111) is true only if $\delta < 0$. In this case, $g(s)$ is an exponentially decaying function of s:

$$Q(s) \sim e^{\delta s^{\gamma+1}} \quad \Rightarrow \quad g(s) \sim s^{-2\beta} e^{\delta s^{\gamma+1}}, \quad s \to \infty \qquad (\delta < 0). \tag{6.114}$$

Chapter 7
Firm's Sudden Deaths

There are a priori two exit mechanisms for firms:

- Firms disappear when their asset values become smaller than some minimum level. This is based on the standard idea, justified by the existence of a minimum efficient size, that there is a minimum firm size below which the firm cannot exist. This idea has been considered in several models of firm growth (see, e.g., de Wit, 2005 and references therein). An alternative approach suggested for instance by Gabaix (1999), considers that firms cannot decline below a minimum size and remain in business at this size until they start growing up again.
- In addition to the exit of a firm resulting from its value decreasing below a certain level, it sometimes happens that a firm encounters financial troubles while its asset value is still fairly high. One could cite the striking examples of Enron Corp. and Worldcom, whose market capitalization were supposedly high (actually the result of inflated total asset value of about $11 billion for Worldcom and probably much higher for Enron) when they went bankrupt. Beyond these anecdotic examples, there is a large empirical literature on firm entries and exits, that suggests the need for taking into account the existence of failure of large firms. For example, while it has been established that a first-order characterization for firm death involves lower failure rates for larger firms (Dunne et al., 1988, 1989), Bartelsman et al. (2003) also state that, for sufficiently old firms, there seems to be no difference in the firm failure rate across size categories.

In previous chapters, we have examined the consequences and impact on Zipf's law of the first exit mechanism. The present chapter is devoted to the study of the second mechanism.

7.1 Definition of the Survival Function

The abrupt death of firms can be described in terms of the random duration T of a firm's life, which is assumed here to be statistically independent from the firm's asset value. We will relax this assumption in the last section of this chapter.

A. Saichev et al., *Theory of Zipf's Law and Beyond*,
Lecture Notes in Economics and Mathematical Systems 632,
DOI 10.1007/978-3-642-02946-2_7, © Springer-Verlag Berlin Heidelberg 2010

If a firm was born at $t = 0$, one can interpret a firm's finite lifespan T as the instant of sudden death of the firm. Taking into account the possible occurrence of such sudden firm's exit, we obtain the mean density $g(s,t)$ of the current firm's asset values, by replacing the pdf $f(s;t)$ in the r.h.s. of expression (3.16) by the product

$$f(s;t)\,\mathcal{Q}(t)\,, \tag{7.1}$$

where $\mathcal{Q}(t)$ is a the survival function, equal by definition to the probability that the life-span T is larger than the current elapsed time t since birth:

$$\mathcal{Q}(t) := \Pr\{T > t\}\,. \tag{7.2}$$

As a result, we obtain the following expression for the mean density of firm's asset values:

Proposition 7.1.1. *Under the assumptions of Corollary 3.2.1 and when firms have finite life durations T independent from their sizes, distributed according to the survival function $\mathcal{Q}(t)$, the mean density of firm's size for homogeneously stochastically growing firms reads*

$$g(s,t) = \int_0^t \nu(t-u)\mathcal{Q}(u)f(s;u)du\,, \tag{7.3}$$

if the first firm was born at time $t_0 = 0$.

This results generalizes Corollary 3.2.2. The intensity of birth $\nu(t)$ can be a priori an arbitrary function of time. Note that, if the pdf $f(s;t)$ takes into account the first possibility of firm's exits due to the crossing of their wealth of some level s_1, the mean density (7.3) then takes into account both possibilities of firm's death listed at the beginning of this chapter.

7.2 Exponential Distribution of Sudden Deaths

Let us now suppose for simplicity that the intensity $\nu(t)$ of firm's births is constant, so that expression (7.3), in the limit $t \to \infty$, reduces to

$$g(s) = \nu \int_0^\infty \mathcal{Q}(t)f(s;t)dt. \tag{7.4}$$

We discuss in appendix the conditions for the existence of this steady-state mean density, which are mush broader than those stated in Sect. 3.3.

In order to account for the sudden exit of firms due to an unexpected event, we introduce the hazard rate μ characterizing the probability that a firm exits in the next instant given it is still alive at the present time. Alternatively, the probability that a firm has never encountered such an event until time t (given it was born at time zero) is equal to the survival function which takes the form of a pure exponential

$$\mathcal{Q}(t) = e^{-\mu t} , \qquad \mu > 0 . \tag{7.5}$$

Thus, we have

$$g(s) = \nu \int_0^\infty e^{-\mu t} f(s;t) dt. \tag{7.6}$$

The relation (6.49) forms the basis of our following analysis of the properties of this steady-state mean density of firm's sizes. Substituting it into the integral (7.6), we obtain, analogously to (6.51), the relation

$$g(s) = \frac{\nu}{b(s)} \eta\left(z(s), \mu\right) , \tag{7.7}$$

where

$$\eta(z, \mu) := \int_0^\infty e^{-\mu t} h(z;t) dt \tag{7.8}$$

and $h(z;t)$ is the solution to the initial-boundary problem (6.43), (6.46) and (6.48).

Multiplying the diffusion equation (6.43) by $e^{-\mu t}$ and integrating it term by term over the interval $t \in [0, \infty)$, we obtain, analogously to (6.54), the ordinary differential equation for the auxiliary function $\eta(z, \mu)$

$$\frac{1}{2} \frac{d^2 \eta(z, \mu)}{dz^2} - \frac{d[c(z)\eta(z, \mu)]}{dz} - \mu \eta(z, \mu) = -\delta(z - z_0) . \tag{7.9}$$

We supplement this equation with the boundary condition

$$\eta(0, \mu) = 0 . \tag{7.10}$$

One might also find another boundary condition adapted to other z-dependences of $c(z)$, by analyzing the properties of some corresponding sample solutions.

7.3 Implications of the Existence of Sudden Firm Exits for Semi-Geometric Brownian Motions

The properties of the steady-state mean density (7.6) of firm's values for the case where the firm's asset values follow semi-GBMs are interesting for economic applications. When the dynamics of the firm's value follows a semi-GBM, the auxiliary pdf $h(z;t)$ is given by the explicit expression (6.66), where c is given by (6.81). Substituting $h(z;t)$ given by (6.66) into (7.8) and using the integral identity (3.23), we obtain an explicit expression for $\eta(z, \mu)$:

$$\eta(z, \mu) = \frac{e^{c(z-z_0)}}{\sqrt{c^2 + 2\mu}} \left(e^{-\sqrt{c^2+2\mu}\,|z-z_0|} - e^{-\sqrt{c^2+2\mu}\,(z+z_0)} \right) . \tag{7.11}$$

Substituting this expression, together with the functions $z(s)$ given by (6.80) and $b(s)$ given by (6.78), into the r.h.s. of relation (7.7), we finally obtain

Proposition 7.3.1. *Under assumptions 1-3 articulated in chapter 3. When the stochastic behavior of firm's sizes is described by the semi-GBM and firms have a finite life duration T (independent from the firm's size) with survival function $\mathcal{Q}(t) = e^{-\mu t}$, the steady-state mean density of firm's sizes is*

$$g(s) = \frac{\nu}{b^2\rho(s+\kappa)} \left(\frac{s+\kappa}{s_0+\kappa}\right)^{\frac{1}{2}(\lambda-\rho)} \times \begin{cases} \left(\frac{s+\kappa}{s_0+\kappa}\right)^{\rho} - \left(\frac{s_1+\kappa}{s_0+\kappa}\right)^{\rho}, & s_1 < s < s_0 , \\ 1 - \left(\frac{s_1+\kappa}{s_0+\kappa}\right)^{\rho}, & s > s_0 , \end{cases} \tag{7.12}$$

where the following notations are used

$$\rho := \sqrt{\lambda^2 + 4\zeta}\,, \qquad \zeta := \frac{2\mu}{b^2}\,. \tag{7.13}$$

Expression (7.12) takes into account the two mechanisms for firm's exits: (1) upon shrinking to a value below the survival threshold s_1 ($0 \leqslant s_1 < s_0$) and (2) abrupt death occurring independently of their value.

Expression (7.12) depends on three characteristic scales of the firm's asset values. The first scale is the initial asset value s_0. The two other scales, s_1 and κ, characterize the deviations from Gibrat's law. The scale s_1 introduces a deviation from Gibrat's law due to death below the survival level which is s_1 itself. The scale κ describes an explicit deviation from Gibrat's law for low firm's values.

In the case where $s_1 = 0$ and when Gibrat's law exactly holds ($\kappa = 0$), expression (7.12) reduces to

$$g(s) = \frac{\nu}{b^2\rho s} \begin{cases} \left(\dfrac{s}{s_0}\right)^{\frac{1}{2}(\lambda+\rho)}, & 0 < s < s_0 , \\ \left(\dfrac{s}{s_0}\right)^{\frac{1}{2}(\lambda-\rho)}, & s > s_0 . \end{cases} \tag{7.14}$$

This result illustrates that, even in the presence of sudden deaths due to external events, for any $s > s_0$, the steady-state mean density $g(s)$ of firm's sizes follows a pure power law

$$g(s) \sim s^{-(1+m)}\,, \qquad s > s_0\,, \tag{7.15}$$

where

$$m = \frac{1}{2}(\rho - \lambda) = \frac{1}{2}\left(\sqrt{\lambda^2 + 4\zeta} - \lambda\right). \tag{7.16}$$

The same power law is obtained from expression (7.12) in the asymptotic limit $s \gg \kappa$ (and for $s > s_0$), showing that it is a robust property also of semi-GBMs.

The asymptotic power law distribution of firm's sizes obtained from (7.12), with tail index given by (7.16), still holds when taking into account that the initial sizes of firms are not identical, but are randomly drawn from some statistical distribution. In this case, it is easy to show that expression (7.6) is modified by taking the expectation of its r.h.s. with respect to the distribution of initial sizes. This yields

$$g(s) \sim \frac{\nu}{b^2 \rho\, s^{1+m}} \left(\mathrm{E}\left[(s_0+\kappa)^m\right] - (s_1+\kappa)^m + o(s)\right) \qquad \text{as } s \to \infty\,. \quad (7.17)$$

provided that $\mathrm{E}\left[s_0^m\right] < \infty$. This generalizes the result stated at the end of Sect. 5.8.

7.4 Zipf's Law in the Presence of Sudden Deaths

The introduction of the mechanism by which firms can exit due to sudden events have a significant influence on the exponent of the power law (7.15). In particular, in the presence of sudden deaths,

Proposition 7.4.1 (Generalized balance condition). *Under the assumptions of Proposition 7.3.1, the condition for Zipf's law to be valid takes the form*

$$\lambda = \zeta - 1 \qquad \Longleftrightarrow \qquad a = \mu\,. \tag{7.18}$$

This condition is a natural generalization of the balance condition $a = 0$ provided by Definition 2.5.1, which was required for Zipf's law to hold for the steady-state mean density $g(s)$ given by (3.18), in the case of a constant intensity of birth and in the absence of sudden death:

$$\nu(t) = \text{const.} \qquad \mathcal{Q}(t) = 1 \qquad \Longrightarrow \qquad \mu = \zeta = 0\,. \tag{7.19}$$

Let us discuss some peculiarities of Zipf's law for the example of semi-GBMs in the presence of sudden deaths together with firm's exits when the value reaches the zero level ($s_1 = 0$). In this case, if condition (7.18) holds, the steady-state mean density $g(s)$ given by Proposition 7.3.1 reduces to

$$g(s) = \frac{\mathcal{N}}{\kappa\left(\frac{s}{\kappa}+1\right)^2\left(\frac{s_0}{\kappa}+1\right)^{\lambda+1}} \begin{cases} \left(\dfrac{s}{\kappa}+1\right)^{\lambda+2} - 1\,, & 0 < \dfrac{s}{\kappa} < \dfrac{s_0}{\kappa}\,, \\ \left(\dfrac{s_0}{\kappa}+1\right)^{\lambda+2} - 1\,, & \dfrac{s}{\kappa} \geq \dfrac{s_0}{\kappa}\,, \end{cases} \tag{7.20}$$

where we have introduced the constant

$$\mathcal{N} := \frac{\nu}{(\lambda+2)b^2}\,. \tag{7.21}$$

The properties of the steady-state mean density $g(s)$ of firm's sizes given by (7.20) depend in an essential way on the ratio of the characteristic scales s_0 and κ. In particular, it follows from (7.20) that, if $\kappa \gg s_0$, then Zipf's law holds only if $s \gg s_0$ (if $s \gtrsim \kappa$). In the opposite case, $\kappa \ll s_0$, Zipf's law is true for any $s > s_0$.

Expression (7.20) shows that the presence of Zipf's law for $s > s_0$ is accompanied by lower power laws over the interval $s \in (0, s_0)$. Their exponents depend on the interplay between the scales κ, s_0 and on the value of the dimensionless parameter λ. Namely, if $\kappa \gg s_0$, then

$$g(s) \simeq (\lambda + 2)\frac{s}{s_0}\,, \qquad s < s_0, \qquad \kappa \gg s_0\,. \tag{7.22}$$

For the opposite case $\kappa \ll s_0$, there are two power laws for the lower values of s. For $s \ll \kappa$, the linear law (7.22) is valid, while if $\kappa \ll s < s_0$, we get the following asymptotic

$$g(s) \simeq \frac{\mathcal{N}}{s_0}\left(\frac{s}{s_0}\right)^{\lambda}, \qquad \kappa \ll s < s_0\,. \tag{7.23}$$

Figures 7.1 and 7.2 plot the steady-state mean density of firm's sizes given by (7.20) for different values of λ and for different ratios of the scales s_0 and κ, inherent to the stochastic behavior of firm's asset values.

In conclusion, consider the properties of the steady-state mean density $g(s)$ for the limiting case $\kappa = 0$, so that, as long as a sudden death does not occur, the stochastic behavior of a firm is governed by the pure Gibrat's law. Supposing that condition (7.18) is valid, we obtain

$$g(s) = \frac{\mathcal{N}}{s_0}\begin{cases}\left(\dfrac{s}{s_0}\right)^{\lambda}, & 0 < s < s_0\,,\\[2ex] \left(\dfrac{s_0}{s}\right)^{2}, & s > s_0\,.\end{cases} \tag{7.24}$$

Let us assume that the initial asset value s_0 is random and is distributed according to the pdf $\gamma(s)$. Averaging the expression (7.24) over the distribution of s_0, we obtain

$$g(s) = \mathcal{N}\left[s^{\lambda}\int_s^{\infty} u^{-(\lambda+1)}\gamma(u)du + s^{-2}\int_0^s u\gamma(u)du\right]. \tag{7.25}$$

Using simple manipulations, we transform the previous relation into

$$g(s) = \mathcal{N}\left[s^{\lambda}\mathcal{F}(s,\lambda) - s^{-1}\mathcal{F}(s,-1) + s^{-2}\int_0^s \mathcal{F}(u,-1)du\right], \tag{7.26}$$

where we have introduced the notation

$$\mathcal{F}(s,\lambda) = \int_s^{\infty} u^{-(\lambda+1)}\gamma(u)du\,. \tag{7.27}$$

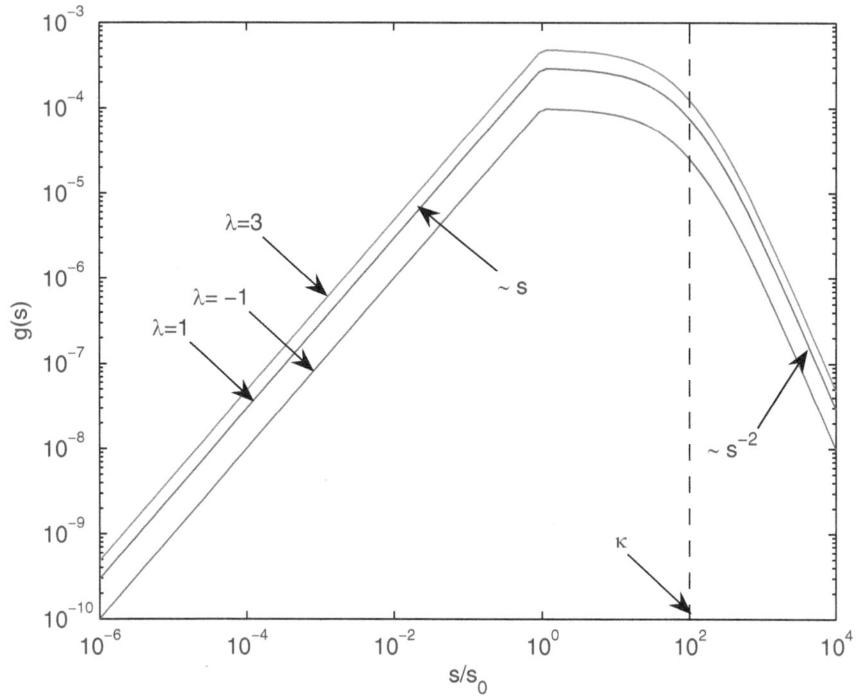

Fig. 7.1 Log-log plots of the steady-state mean density $g(s)$ given by (7.20) of firm's sizes for $\kappa = 100 s_0$ and for $\mathcal{N} = 1$. *From bottom to top,* $\lambda = -1, 1, 3$. The linear behavior (7.22) is clearly visible for the interval $s \in (0, s_0)$, and Zipf's law can be observed for $s \gtrsim \kappa$

Let $\gamma(s)$ be the gamma distribution given by (6.88), then

$$\mathcal{F}(s,\lambda) = \mathcal{F}(s,\bar{s},q,\lambda) = \left(\frac{q}{\bar{s}}\right)^{\lambda+1} \frac{\Gamma\left(q-\lambda-1, qs/\bar{s}\right)}{\Gamma(q)} . \tag{7.28}$$

Furthermore, the integral in the last term of equality (7.26) is described by expression (6.89). Figure 7.3 plots the steady-state mean density of firm's sizes given by (7.26) for different values of λ.

7.5 Explanation of the Generalized Balance Condition

We now discuss the meaning of the balance condition given by Proposition 7.4.1 for the validity of Zipf's law. In this goal, let us first consider the case where sudden deaths are absent, for which the natural balance condition introduced by Definition 2.5.1, i.e., $\lambda = -1$ or equivalently $a = 0$, is necessary for Zipf's law to hold. Assuming that Gibrat's law holds, i.e., the stochastic behavior of the firm's asset

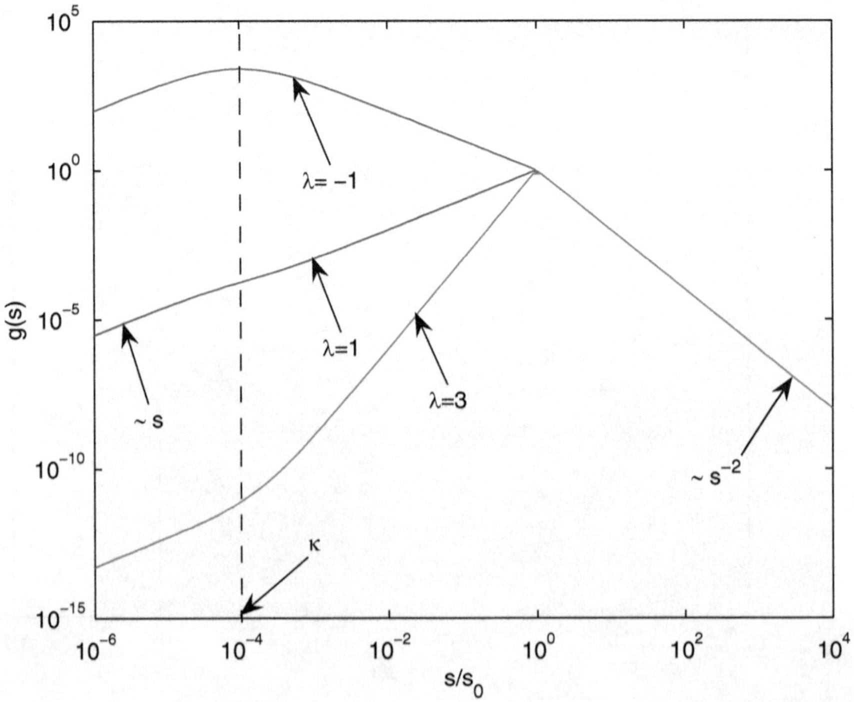

Fig. 7.2 Log-log plots of the steady-state mean density $g(s)$ of firm's sizes given by (7.20) for $\kappa = 10^{-4}s_0$ and $\mathcal{N} = 1$. *From top to bottom* $\lambda = -1$, 1, 3. Zipf's law is seen for $s > s_0$ and the crossover between linear and power laws (7.22), (7.23) can be observed within the interval $s \in (0, s_0)$

values $S(t)$ is described by GBMs, then the statistical average of $S(t)$ is equal to

$$E[S(t)] = s_0\, e^{at}\,. \tag{7.29}$$

In this case, the balance condition $a = 0$ means that the statistical average of firm's asset values does not depend on time:

$$E[S(t)] = s_0 \quad \Longleftrightarrow \quad a = 0 \quad \Longleftrightarrow \quad \lambda = -1\,. \tag{7.30}$$

In other words, the balance condition means that the overall asset values of firms that were born at the same time is constant.

Consider now the statistical average of some firm's asset value $E_d[S(t)]$, where the index d indicates that this average takes into account the possibility of the firm's sudden death. Obviously, this average is equal to

$$E_d[S(t)] = \mathcal{Q}(t)E[S(t)]\,, \tag{7.31}$$

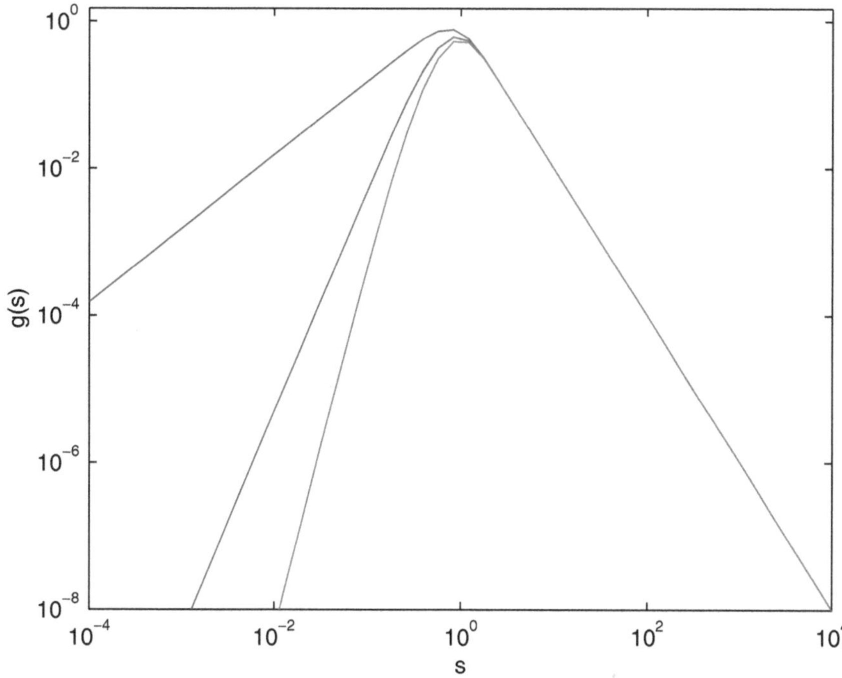

Fig. 7.3 Plots of the steady-state mean density given by (7.26) for $\bar{s} = 1$ and $q = 8$. *From top to bottom,* $\lambda = 1, 3, 5$

where $E[S(t)]$ is the statistical average of the firm's asset value $S(t)$ in the absence of sudden death and $\mathcal{Q}(t)$ is a the survival function defined in Sect. 7.1. Assuming for simplicity that $S(t)$ is a GBM whose average is given by relation (7.29), while the survival function in the presence of sudden deaths is the exponential (7.5), we find that the average firm value $E_d[S(t)]$, taking into account the existence of sudden deaths and for a GBM dynamics of the firm's asset values before death, is equal to

$$E_d[S(t)] = s_0 \, e^{(a-\mu)t} \, . \tag{7.32}$$

If condition (7.18) is valid, then we have

$$E_d[S(t)] = s_0 \, . \tag{7.33}$$

Thus, condition (7.18) necessary for Zipf's law to be valid is nothing but a generalization of the Definition 2.5.1 of a balanced growth for the general case $\mu \neq 0$. In other words, if condition (7.18) holds, then in spite of the existence of sudden deaths, the overall asset value of surviving firms that were born at the same time does not depend on time and remains equal to the overall initial asset value.

7.6 Some Consequences of the Generalized Balance Condition

Let us now consider in detail the average $E_d[S(t)]$ given by (7.31), in which the factor $E[S(t)]$ in the r.h.s. of relation (7.31) takes into account both the killing of firms upon first reaching the survival level $s_1 \geqslant 0$ and deviations from Gibrat's law due to a non-zero $\kappa > 0$. Obviously

$$E[S(t)] = \int_{s_1}^{\infty} s f(s;t) ds, \tag{7.34}$$

where $f(s;t)$ is the solution to the diffusion equation (2.39), satisfying both initial (2.40) and boundary (6.48) conditions. Using relations (6.41) and (6.49), we rewrite the last equality in the form

$$E[S(t)] = \int_{0}^{\infty} s(z) h(z;t) dz\,, \tag{7.35}$$

where $h(z;t)$ is the solution to the initial-boundary problem (6.43), (6.46), (6.48). For the particular case where $S(t)$ follows a semi-GBM, for which the pdf $h(z;t)$ is given by the explicit expression (6.66), (6.67), we obtain

$$E[S(t)] = (s_1 + \kappa)\mathcal{E}(t) - \kappa\mathcal{E}_0(t)\,, \tag{7.36}$$

where

$$\mathcal{E}(t) = \int_{0}^{\infty} e^{bz} h(z;t) dz\,, \qquad \mathcal{E}_0(t) = \int_{0}^{\infty} h(z;t) dz\,. \tag{7.37}$$

After simple calculations, one obtains

$$\begin{aligned}\mathcal{E}(t) = \frac{1}{2} e^{(1+\lambda)\tau} \Bigg[\frac{s_0+\kappa}{s_1+\kappa} \operatorname{erfc}\left(\frac{1}{2\sqrt{\tau}} \ln\left[\frac{s_1+\kappa}{s_0+\kappa}\, e^{-(\lambda+2)\tau} \right] \right) \\ - \left(\frac{s_1+\kappa}{s_0+\kappa} \right)^{\lambda+1} \operatorname{erfc}\left(\frac{1}{2\sqrt{\tau}} \ln\left[\frac{s_0+\kappa}{s_1+\kappa}\, e^{-(\lambda+2)\tau} \right] \right) \Bigg]\end{aligned} \tag{7.38}$$

and

$$\begin{aligned}\mathcal{E}_0(t) = \frac{1}{2} \Bigg[\operatorname{erfc}\left(\frac{1}{2\sqrt{\tau}} \ln\left[\frac{s_1+\kappa}{s_0+\kappa}\, e^{-\lambda\tau} \right] \right) \\ - \left(\frac{s_1+\kappa}{s_0+\kappa} \right)^{\lambda} \operatorname{erfc}\left(\frac{1}{2\sqrt{\tau}} \ln\left[\frac{s_0+\kappa}{s_1+\kappa}\, e^{-\lambda\tau} \right] \right) \Bigg] .\end{aligned} \tag{7.39}$$

In particular, it follows from these results that, if condition (7.18) is valid, then there is a finite non-zero limit

$$\lim_{\tau\to\infty} E_d[S(t)] = \lim_{\tau\to\infty} E[S(t)] e^{-(1+\lambda)\tau} = s_0\, \Omega(x, y, \lambda)\,, \tag{7.40}$$

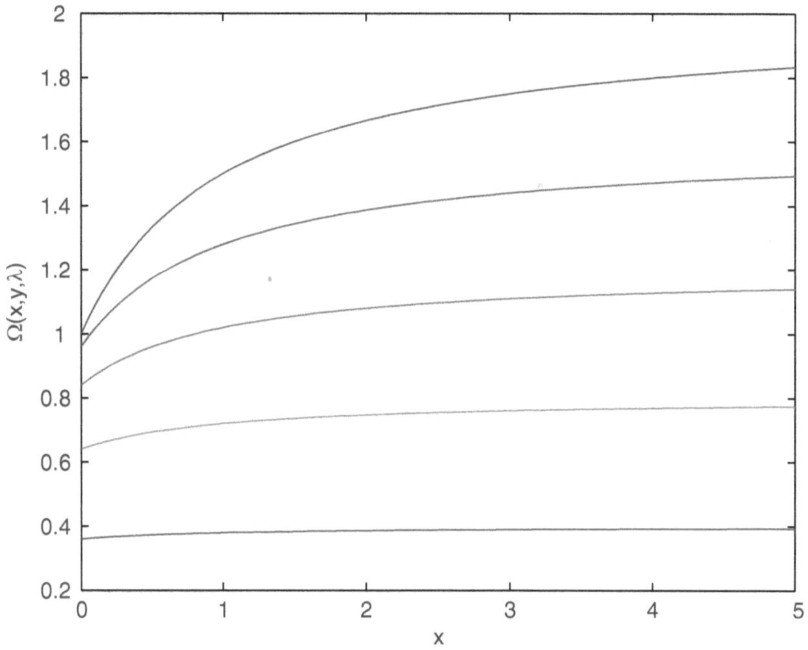

Fig. 7.4 Plot of the ratio $\Omega(x, y, \lambda)$ given by expression (7.41) as a function of x for $\lambda = 0$. *From top to bottom,* $y = 0, 0.2, 0.4, 0.6, 0.8$. It can be seen that the average $E[S(\infty)]$ of the asset values of the surviving firms grows larger with κ, and may be larger than the initial asset value s_0

where

$$\Omega(x, y, \lambda) = (1 + x)\left[1 - \left(\frac{y + x}{1 + x}\right)^{\lambda+2}\right], \quad x = \frac{\kappa}{s_0}, \quad y = \frac{s_1}{s_0}, \quad \lambda > -1 . \tag{7.41}$$

One can interpret Ω as the ratio of the long-term average asset value $E[S(\infty)]$ of surviving firms to the initial asset value s_0. The finiteness of the limit (7.40) means that, despite sudden deaths and killing at level s_1, the mean asset value of surviving firms remains finite and comparable with the initial value s_0. In this sense, condition (7.18) is actually a balance condition, even in the presence of firms being killed at level s_1. Figure 7.4 shows the function $\Omega(x, y, \lambda)$.

7.7 Zipf's Law as a Universal Law with a Large Basin of Attraction

We have shown that, for given parameters a, μ, and volatility b, the steady-state mean density of firm's asset values has the power asymptotics

$$g \sim s^{-(m+1)} , \tag{7.42}$$

where the exponent m given by (7.16) is a function of the values of the parameters a, μ and b. In particular, for different coefficients of volatility b, and unchanged parameters a and μ, the exponent of the power law (7.42) is different.

There is a unique exception to this behavior, which occurs when Zipf's law holds. Indeed, if the general balance condition (7.18) is valid, then for any b we have $m = 1$. In contrast, if $a \neq \mu$, then it is not possible to find a value for the volatility $b < \infty$ giving $m = 1$. In other words, Zipf's law exhibits a universal behavior in the sense that, if it is true for some volatility b_1, then it will be true for any other volatility b_2.

At the same time, as the volatility b increases, the power law (7.42) becomes closer to Zipf's law, and the later is obtained asymptotically for very large volatilities b for all values of the other parameters. In this sense, Zipf's law is "attracting" the power laws (7.42). To illustrate this property of Zipf's law, we rewrite the expression of the exponent (7.16) in the form

$$m = \frac{1}{2\delta'} \left(\sqrt{(\delta' - 1)^2 + 4\delta' \varepsilon} + \delta' - 1 \right) , \tag{7.43}$$

where the distance of the ratio

$$\varepsilon := \frac{\mu}{a} \tag{7.44}$$

to 1 provides a quantification of how much the condition (7.18) is violated. The parameter

$$\delta' := \frac{1}{1 + \lambda} = \frac{b^2}{2a} \tag{7.45}$$

is proportional to b^2 and goes to infinity as $\lambda \to -1$, i.e., approaches the balance condition in absence of sudden death. Plots of the exponent m given by (7.43) as a function of δ' are depicted in Fig. 7.5, for different values of the ratio ε. It can be seen in particular that the greater δ', i.e., the more volatile are the GBMs of the asset values, the closer the exponent m is to 1, and the closer the power law (7.42) is to Zipf's law.

7.8 Rate of Sudden Death Depending on Firm's Asset Value

Until now, we have considered only the case where the abrupt death of a given firm was independent of the firm's asset value, and was controlled by a simple stationary process. Accordingly, the survival function defined in Sect. 7.1 was a pure exponential with constant hazard coefficient μ.

We now investigate the more general case where the hazard coefficient is a function of the firm's asset value and of the age of the firm. This implies that the survival function $\mathcal{Q}(t)$ of a firm which was born at the instant $t = 0$ satisfies the general equation

$$d\mathcal{Q}(t) = -\mu\left(S(t), t\right) \mathcal{Q}(t)\, dt \,, \qquad \mathcal{Q}(0) = 1 \,. \tag{7.46}$$

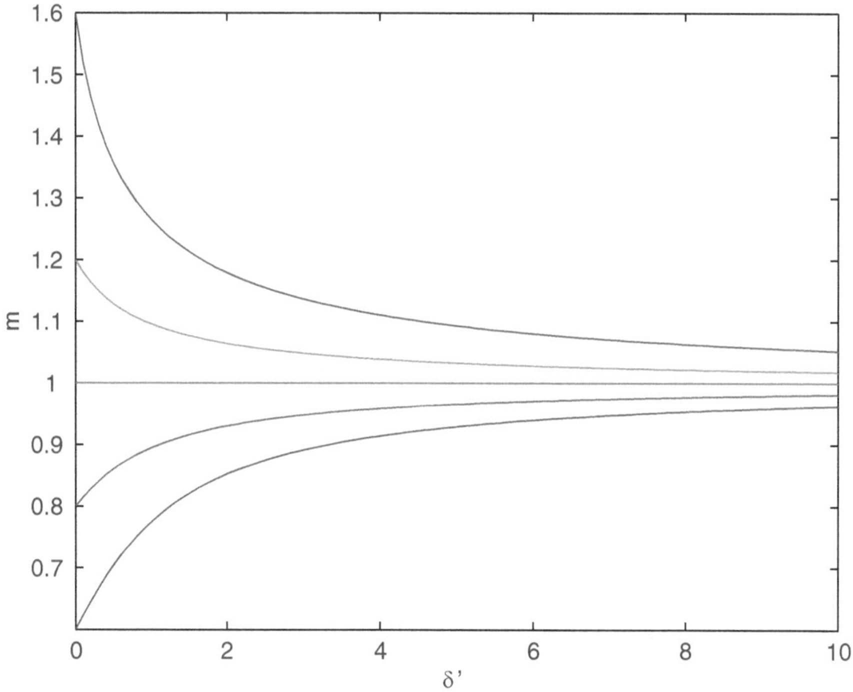

Fig. 7.5 Plot of the dependence of the exponent m given by (7.43) as a function of δ' defined by (7.45) for different fixed values of the ratio $\varepsilon = \mu/a$. *From bottom to top,* $\varepsilon = 0.6, 0.8, 1, 1.2, 1.4$. The greater the volatility is, the closer the power law (7.42) is to Zipf's law

Its solution reads

$$\mathcal{Q}(t) = \exp\left(-\int_0^t \mu(S(t'), t')dt'\right) . \tag{7.47}$$

Note that, as $S(t)$ is in general described by a stochastic process, then the survival function $\mathcal{Q}(t)$ is also random. The corresponding pdf of the firm's asset value is given by

$$f(s;t) = E\left[\mathcal{Q}(t)\delta(S(t) - s)\right] . \tag{7.48}$$

Let us assume that $S(t)$ satisfies the stochastic equation (6.1). It is easy to show that in this case the pdf $f(s;t)$ given by (7.48) satisfies the equation

$$\frac{\partial f(s;t)}{\partial t} + \frac{\partial [a(s)f(s;t)]}{\partial s} + \mu(s,t)f(s;t) = \frac{1}{2}\frac{\partial^2[b^2(s)f(s;t)]}{\partial s^2} . \tag{7.49}$$

with the initial condition (2.40). Correspondingly, if the intensity of birth is constant, while the hazard rate does not depend on the age of the firm, i.e., if $\mu = \mu(s)$, then the steady-state mean density $g(s)$ defined in (3.18) satisfies the equation

$$\frac{1}{2}\frac{d^2[b^2(s)g(s)]}{ds^2} - \frac{d[a(s)g(s)]}{ds} - \mu(s)g(s) = -\delta(s - s_0) . \tag{7.50}$$

As an example, let us calculate the steady-state mean density $g(s)$ of firm's sizes, supposing that $S(t)$ follows a GBM, such that $a = 0$ and $b(s) = b \cdot s$, while $\mu(s)$ is a linear function of s: $\mu(s) = \mu \cdot s$. In this case, introducing the auxiliary function $\psi(s) := s^2 g(s)$, (7.50) reduces to the homogeneous equation

$$s\frac{d^2\psi(s)}{ds^2} - \zeta\psi(s) = 0\,, \tag{7.51}$$

with $\zeta = \frac{2\mu}{b^2}$ as defined in Proposition 7.3.1. In addition, the function $\psi(s)$ has to obey the following conditions of continuity at s_0

$$\psi(s_0 - 0) = \psi(s_0 + 0)\,, \qquad \frac{d^2\psi(s)}{ds^2}\Big|_{s_0+0}^{s_0-0} = \frac{2}{b^2}\,, \tag{7.52}$$

in order to take into account the delta function in the r.h.s. of (7.50). A general solution to (7.51), expressed in terms of the steady-state mean density

$$g(s) = \psi(s)/s^2\,, \tag{7.53}$$

reads

$$g(s) = \frac{1}{s}\sqrt{\frac{\zeta}{s}}\left[A\, I_1\left(2\sqrt{\zeta s}\right) + B\, K_1\left(2\sqrt{\zeta s}\right)\right]\,, \tag{7.54}$$

where $I_1(x)$ and $K_1(x)$ are respectively the modified Bessel functions of the first and second kind. A and B are two constants to be determined.

In order to fully determine $g(s)$ from (7.54), it is convenient to use the inverse relation pointed out in Sect. 2.3 with (2.23). We compare the asymptotic, for $\zeta \to 0$, of the two terms of the general solution (7.54) and the corresponding steady-state mean density given by (3.24), which for $\lambda = -1$ is equal to

$$g(s) = \frac{2\nu}{b^2}\begin{cases} s^{-1}\,, & 0 < s < s_0\,, \\ s_0 s^{-2}\,, & s > s_0\,. \end{cases} \tag{7.55}$$

The following asymptotics are true:

$$\frac{1}{s}\sqrt{\frac{\zeta}{s}} I_1\left(2\sqrt{\zeta s}\right) \simeq \frac{\zeta}{s}\,, \qquad \frac{1}{s}\sqrt{\frac{\zeta}{s}} K_1\left(2\sqrt{\zeta s}\right) \simeq \frac{1}{2s^2}\,, \qquad s \to 0\,. \tag{7.56}$$

It follows from these asymptotics and from the limiting mean density (7.55), that the sought mean density is equal to

$$g(s) = \frac{2\nu}{b^2} C \frac{1}{s}\sqrt{\frac{\zeta}{s}}\begin{cases} I_1\left(2\sqrt{\zeta s}\right)/\zeta\,, & 0 < s < s_0\,, \\ s_0 D K_1\left(2\sqrt{\zeta s}\right)\,, & s > s_0\,, \end{cases} \tag{7.57}$$

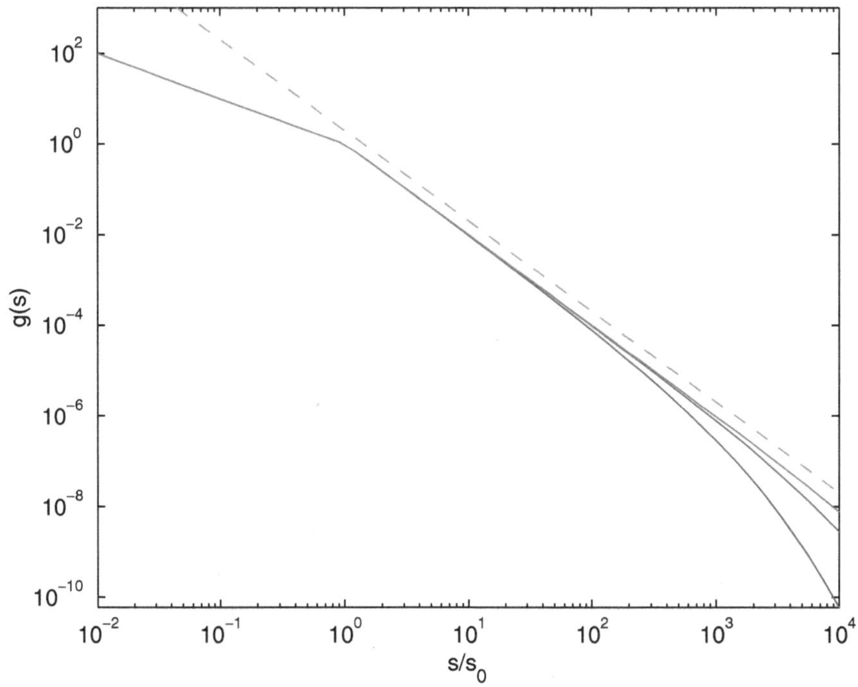

Fig. 7.6 Log-log plot of expression (7.57) for the mean density of firm's sizes, demonstrating the distortion of Zipf's law due to the linear dependence of the hazard coefficient $\mu \cdot s$ on the asset value s. *From bottom to top,* $\zeta = \frac{2\mu}{b^2} = 10^{-3}, 10^{-4}, 10^{-5}$. The *upper dashed straight line* shows the exact Zipf's law $g(s) \sim s^{-2}$

where the two constants C and D are determined as follows. The constant D is determined from the continuity condition (7.52):

$$D = \frac{I_1\left(2\sqrt{\zeta s_0}\right)}{\zeta s_0 K_1\left(2\sqrt{\zeta s_0}\right)} . \tag{7.58}$$

The other proportionality constant C is obtained by using the second jump condition (7.52). Figure 7.6 shows the mean density of firm's sizes given by (7.57), demonstrating the distortion of Zipf's law due to the linear dependence $\mu(s) = \mu \cdot s$ of the hazard coefficient as a function of the asset value s.

7.9 Rate of Sudden Death Depending on Firm's Age

We now focus on the case where the hazard rate μ depends only on the age of the firms. This situation is important insofar as the literature often reports that new born firms have a much larger failure rate than well-established firms (Becchetti and Trovato, 2002, for instance). One of the goals of this section is to analyze possible deviation from Zipf's law of the mean density $g(s)$ of firm's assets resulting from a time-dependent hazard rate.

Following the same approach as previously, we explore the properties of the mean density

$$g(s) = \nu \int_0^{\infty} \mathcal{Q}(u) f(s; u) du \,, \tag{7.59}$$

where the surviving function $\mathcal{Q}(t)$ takes into account the dependence of the surviving probability with respect to the age of the firms. In order to account for the larger failure rate of newly born firms, let us choose

$$\mathcal{Q}(t) = \exp\left(- \int_0^t \mu(t') dt' \right) \,, \tag{7.60}$$

with

$$\mu(t) = \mu_\infty + (\mu_0 - \mu_\infty) e^{-kt} \,. \tag{7.61}$$

Here, k is a time scale factor. Instead of e^{-kt}, we could have chosen any arbitrary monotonically decreasing function, such that $h(0) = 1$ and $h(\infty) = 0$. However, the present choice allows us to obtain analytical results. In this case

$$\mathcal{Q}(t) = \exp\left(-\mu_\infty t - \frac{\mu_0 - \mu_\infty}{k} \left(1 - e^{-kt}\right) \right) \,. \tag{7.62}$$

It is convenient to rewrite this expression in dimensionless form, taking as the natural time scale the characteristic time $t_b = 2/b^2$ given by (2.18), associated with the volatility of the GBM. Expression (7.62) thus becomes

$$\mathcal{Q}(t) = \exp\left(-\zeta_\infty \tau - \frac{\zeta_0 - \zeta_\infty}{\kappa} \left(1 - e^{-\kappa\tau}\right) \right) \,, \tag{7.63}$$

where

$$\zeta_0 := \frac{2\mu_0}{b^2} \,, \qquad \zeta_\infty := \frac{2\mu_\infty}{b^2} \,, \qquad \kappa := \frac{2k}{b^2} \,. \tag{7.64}$$

The parameter κ, in a unit inverse to the volatility time t_b given by (2.18), is the rate of switching between the two asymptotic exponential surviving regimes:

$$\mathcal{Q}(t) \simeq e^{-\mu_0 t} = e^{-\zeta_0 \tau} \quad (\kappa\tau \ll 1) \qquad \mathcal{Q}(t) \sim e^{-\mu_\infty t} = e^{-\zeta_\infty \tau} \quad (\kappa\tau \gg 1) \tag{7.65}$$

Let us determine the order of magnitude of the different parameters. First, $t_b = 2/b^2$ is the characteristic volatility time. For an annual volatility $b \simeq 0.2$, we obtain $t_b \simeq 50$ years. It seems reasonable to consider that the formation of an "adult" firm takes approximately $1/k \simeq 1 \sim 5$ years, leading to a characteristic value of the parameter κ (7.64) in the range $\kappa \simeq 10 \sim 50$. It also appears reasonable to take as typical that the characteristic lifetime of an "adult" firm is approximately 25 years[1]

$$\frac{1}{\mu_\infty} \simeq \frac{t_b}{2} \simeq 25 \text{ years} \quad \Rightarrow \quad \zeta_\infty \simeq 2\,. \tag{7.66}$$

In order to describe the fact that young firms are more prone to fail, we assume that the characteristic lifetime of young firms is approximately 5 years:

$$\frac{1}{\mu_0} \simeq \frac{t_b}{10} \simeq 5 \text{ years} \quad \Rightarrow \quad \zeta_0 \simeq 10\,. \tag{7.67}$$

For a numerical illustration of the effect of a time-dependent hazard rate, this suggests to choose the parameters defined in (7.64) as follows:

$$\zeta_\infty = 2\,, \qquad \zeta_0 = 10\,, \qquad \kappa = 10 \sim 50\,. \tag{7.68}$$

Figure 7.7 shows a log-log plot of the surviving function (7.63), for the parameter values given by (7.68), for different values of κ.

Substituting in (7.59) the log-normal pdf $f(s;t)$ given by (4.62) and the survival function (7.60), we obtain

$$g(s) = \frac{\bar{\nu}}{s} \int_0^\infty \frac{d\tau}{\sqrt{\pi\tau}} \times \exp\left(-\zeta_\infty \tau - \frac{\zeta_0 - \zeta_\infty}{\kappa}\left(1 - e^{-\kappa\tau}\right) - \frac{\left(\ln\left(\frac{s}{s_0}\right) + \tau - \delta\tau\right)^2}{4\tau}\right). \tag{7.69}$$

where $\bar{\nu} := \frac{\nu}{b^2}$ is the dimensionless rate of firm's births.

We are now in position to explore numerically the possible deviations from Zipf's law, caused by a non-exponential form of the survival function $Q(t)$ (7.63) which takes into account different decaying rates for "young" and "adult" firms. In the case of a pure exponential survival function given by (7.5), recall that the balance condition $\lambda = \zeta - 1$ ensures the validity of Zipf's law. Let us here assume that the balance condition holds for "adult" firms, i.e., $\lambda = \zeta_\infty - 1$. For the parameter values (7.68), Fig. 7.8 shows the mean density given by (7.69). One can conclude that the existence of different rates of deaths for "young" and "adult" firms does not modify significantly Zipf's law when the balance condition $\lambda = \zeta_\infty - 1$ for "adult" firms holds.

We can also resort to a saddle point approximation to derive the large s behavior of the distribution of firm's sizes given by (7.69). Let us first rewrite this relation as

$$g(s) = \frac{\bar{\nu}}{s\sqrt{\pi}} \int_0^\infty d\tau\, e^{-f_x(\tau)}\,, \tag{7.70}$$

[1] Daniel et al. (2008) find that the half-time (time needed for half a population to disappear) of all firms recorded on the CRSP database is significantly shorter, of the order of 5–10 years.

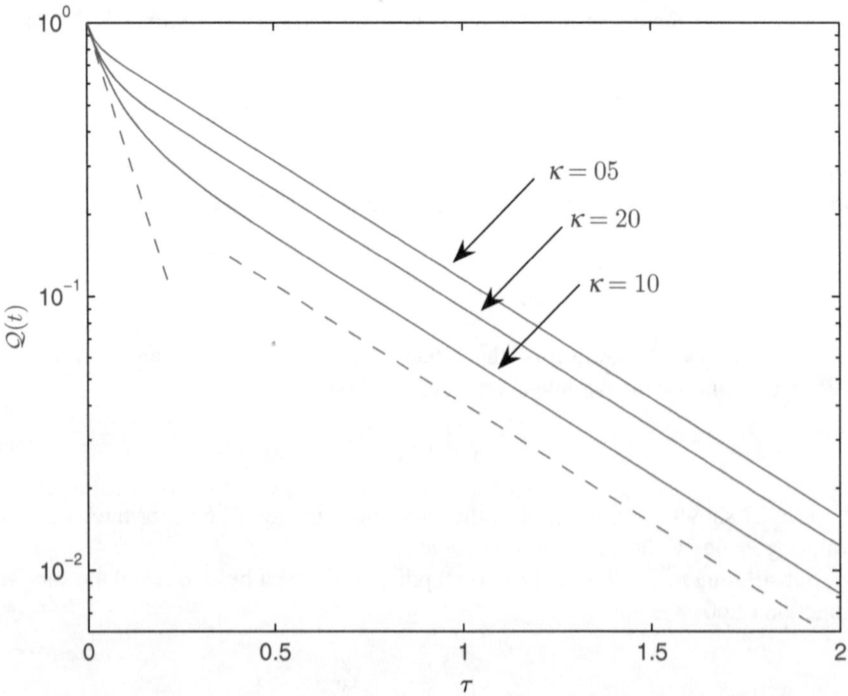

Fig. 7.7 Log-log plot of the surviving function $\mathcal{Q}(t)$ given by (7.63), for the parameters values given by (7.68). The *two dashed lines* are the exponential asymptotes given by (7.65)

with

$$f_x(\tau) = \frac{1}{2}\ln\tau + \zeta_\infty\tau + \frac{\zeta_0 - \zeta_\infty}{\kappa}\left(1 - e^{-\kappa\tau}\right) + \frac{1}{4\tau}\left(x - \lambda\tau\right)^2 \ , \tag{7.71}$$

and x stands for $\ln\left(\frac{s}{s_0}\right)$. Expanding $f_x(\tau)$ around its minimal value τ^*, in the limit of large x, we get

$$f_x(\tau) \approx f_x(\tau^*) + \frac{1}{2}f_x''(\tau^*)(\tau - \tau^*) \tag{7.72}$$

with

$$\tau^* \approx x \cdot \left[\lambda^2 + 4\zeta_\infty\right]^{-1/2} \tag{7.73}$$

and

$$f_x(\tau^*) \approx \frac{1}{2}\ln\frac{x}{\sqrt{\lambda^2 + 4\zeta_\infty}} + \frac{\zeta_0 - \zeta_\infty}{\kappa} + \frac{x}{2}\left[\sqrt{\lambda^2 + 4\zeta_\infty} - \lambda\right], \tag{7.74}$$

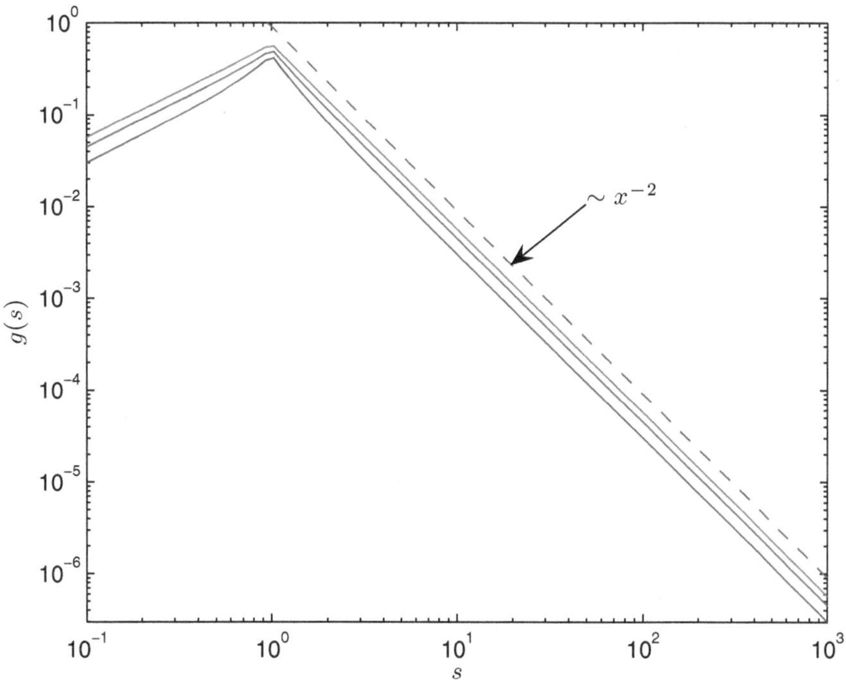

Fig. 7.8 *Solid lines:* log-log plots of the steady-state mean density $g(s)$ given by (7.69) for the parameters $\lambda = 1$, $\zeta_\infty = 2$, $\zeta_0 = 10$, $\kappa = 10 \sim 50$, $s_0 = 1$, $\bar{\nu} = 1$. *Bottom to top:* $\kappa = 10, 20, 50$. The *dashed line* is the pure Zipf's law $g(s) \sim s^{-2}$

$$f_x''(\tau^*) \approx \frac{1}{2x\left[\lambda^2 + 4\zeta_\infty\right]^{3/2}} . \tag{7.75}$$

After integration of the Gaussian term that results from the second order approximation of expression (7.70), we obtain

$$g(s) \approx \frac{2\bar{\nu}\, e^{\frac{\zeta_0 - \zeta_\infty}{\kappa}}}{\sqrt{\lambda^2 + 4\zeta_\infty}} \left(\frac{s}{s_0}\right)^{-(1+m)} \tag{7.76}$$

with

$$m = \frac{1}{2}\left(\sqrt{\lambda^2 + 4\zeta_\infty} - \lambda\right) . \tag{7.77}$$

Thus, provided that the hazard rate of firm's sudden death goes to some constant for large time, the asymptotic behavior of the distribution of firm's size is left unchanged. The tail index of the distribution is still controlled by the parameter λ and the *long term* hazard rate of firm's deaths.

Appendix

Interrelation Between the Steady-State Mean Density $g(s)$ and the Steady-State Pdf $f(s)$

In this appendix, we come back to the discussion started in Sect. 3.3 about the conditions of stationarity of the mean density of firm's sizes. We had seen that, if the integral

$$\nu \int_0^{\infty} f(s;t)dt \tag{7.78}$$

converges so that the corresponding steady-state density $g(s)$ exists, then a nonzero steady-state distribution in Gabaix's sense cannot exist, and vice-verse. Indeed, mathematically speaking, the existence of Gabaix's steady-state distribution is equivalent to the existence of a nonzero limit

$$f(s) = \lim_{t \to \infty} f(s;t) \,, \tag{7.79}$$

i.e., to the existence of a steady-state pdf $f(s)$ of individual asset values. It is obvious that, if $f(s) > 0$ for some s, then the improper integral (7.78) diverges and the steady-state density $g(s)$ does not exist. There is thus a contradiction between the existence of a steady-state density in Gabaix's sense and in the presence of a flow of firm's birth which is central to this book.

Nevertheless, if one takes into account not only the birth flow, but also the possible deaths of firms, then a steady-state mean density $g(s)$ may exist even if the pdf $f(s,t)$ admits a non-zero limit. To explain this fact, consider the improper integral

$$\nu \int_0^{\infty} \mathcal{Q}(u) f(s;u)du \,, \tag{7.80}$$

which is the limit, if it exists, of (7.3) as t goes to infinity, in the case where the mean rate of firms birth ν is constant. As it will be seen hereafter, a monotonically decaying function $\mathcal{Q}(u)$ can describe either the process of firm's death or inhomogeneous mean rates of firm births.

To ensure the consistency between the steady-state mean density $g(s)$ and the steady-state pdf $f(s)$, notice that it is easy to prove the following

Proposition 7.9.1. *If $f(s;u)$ is a bounded function of u, i.e., there exists a positive constant $0 < B(s) < \infty$ such that*

$$f(s;u) < B(s) \qquad \forall\, u \in [0, \infty) \,,$$

while $\mathcal{Q}(u)$ is a function, satisfying the asymptotic relation

$$\mathcal{Q}(u) \sim o\{u^{-p}\} \,, \qquad u \to \infty \,, \qquad p > 1 \,,$$

then the improper integral (7.80) *converges, and the steady-state mean density* $g(s)$ *exists for the range of* s *such that the first condition is fulfilled.*

The above assertion remains true even when $f(s;u)$ presents an integrable singularity at $u \to 0_+$ of the form (3.21)

$$f(s;u) \sim O\left\{\frac{1}{\sqrt{u}}\right\} , \qquad u \to 0_+ .$$

Let us illustrate the joint existence, as $t \to \infty$, of a mean density of firm's sizes $g(s)$ and of a distribution $f(s)$ of individual asset value using the example of the GBM $S(t)$ satisfying the stochastic equation

$$dS(t) = a \cdot S(t)\, dt + b \cdot S(t)\, dW(t) , \qquad S(t=0) = s_0 > 0 , \tag{7.81}$$

where $W(t)$ is a standard Wiener process. The corresponding pdf $f(s;t)$ satisfies the diffusion equation

$$\frac{\partial f(s;t)}{\partial t} + a\frac{\partial s f(s;t)}{\partial s} = \frac{b^2}{2}\frac{\partial s^2 f(s;t)}{\partial s^2} . \tag{7.82}$$

Suppose that a reflecting barrier is located at $s = s_1$, such that the following boundary condition holds (Karatzas and Shreve, 1991)

$$\mathcal{F}(s_1;t) = 0 , \tag{7.83}$$

where

$$\mathcal{F}(s;t) = a\ s\ f(s;t) - \frac{b^2}{2}\frac{\partial s^2 f(s;t)}{\partial s} \tag{7.84}$$

is the flow of the probability measure.

Let us find the steady-state solution to the diffusion equation (7.82), satisfying the boundary condition (7.83). Noticing that the steady-state pdf corresponds to a zero flow for any $s > s_1$, we obtain

$$\lim_{t\to+\infty} \mathcal{F}(s,t) \equiv \mathcal{F}(s) = a\ s\ f(s) - \frac{b^2}{2}\frac{ds^2 f(s)}{ds} = 0 , \qquad s > s_1 . \tag{7.85}$$

A non-vanishing solution of the last equation, satisfying the normalizing condition

$$\int_{s_1}^{\infty} f(s) ds = 1 ,$$

exists for

$$\lambda < 0 . \tag{7.86}$$

It takes the power law form

$$f(s) = \frac{|\lambda|}{s_1}\left(\frac{s}{s_1}\right)^{\lambda-1} \sim s^{\lambda-1}\,, \qquad s > s_1\,, \quad \lambda < 0\,. \tag{7.87}$$

As discussed previously, Zipf's law follows from the balance condition $\lambda = -1$.

We now determine the corresponding steady-state density $g(s)$, assuming that $\mathcal{Q}(t)$ is exponential

$$\mathcal{Q}(t) = e^{-\mu t}\,, \qquad \mu > 0\,. \tag{7.88}$$

For this, we first determine the solution to the diffusion equation (7.82), satisfying the initial condition

$$f(s, t = 0) = \delta(s - s_0) \tag{7.89}$$

and the boundary condition (7.83). By the change of variable

$$z := \frac{1}{b}\ln\left(\frac{s}{s_0}\right) \qquad \Longleftrightarrow \qquad s = s_0\, e^{bz}\,, \tag{7.90}$$

the boundary-initial problem (7.82), (7.83), (7.89) reduces to the auxiliary initial-boundary problem

$$\begin{gathered} \frac{\partial\varphi(z;t)}{\partial t} + \varrho\frac{\partial\varphi(z;t)}{\partial z} = \frac{1}{2}\frac{\partial^2\varphi(z;t)}{\partial z^2}\,, \\ \varphi(z;t=0) = \delta(z)\,, \qquad \mathcal{G}(z_1;t) = 0\,, \end{gathered} \tag{7.91}$$

where $\varphi(z;t)$ and $f(s;t)$ are tied by the relation

$$f(s;t) = \frac{1}{b\cdot s}\varphi\left[\frac{1}{b}\ln\left(\frac{s}{s_0}\right);t\right] \tag{7.92}$$

and

$$\mathcal{G}(z;t) := \varrho\,\varphi(z;t) - \frac{1}{2}\frac{\partial\varphi(z;t)}{\partial z}\,, \quad z_1 := \frac{1}{b}\ln\left(\frac{s_1}{s_0}\right)\,, \quad \varrho := \frac{a}{b} - \frac{b}{2}\,. \tag{7.93}$$

A standard approach to obtain the solution to the initial-boundary problem (7.91) is to use the reflection method (Borodin and Salminen, 2002), which leads to

$$\begin{aligned} \varphi(z;t) =& \frac{1}{\sqrt{2\pi t}}\exp\left(-\frac{(z-\varrho t)^2}{2t}\right) \\ &+ e^{2\varrho z_1}\frac{1}{\sqrt{2\pi t}}\exp\left(-\frac{(z-2z_1-\varrho t)^2}{2t}\right) \\ &- \varrho e^{2\varrho(z-z_1)}\operatorname{erfc}\left(\frac{z-2z_1+\varrho t}{\sqrt{2t}}\right)\,, \qquad z > z_1,\ \ z_1 < 0. \end{aligned} \tag{7.94}$$

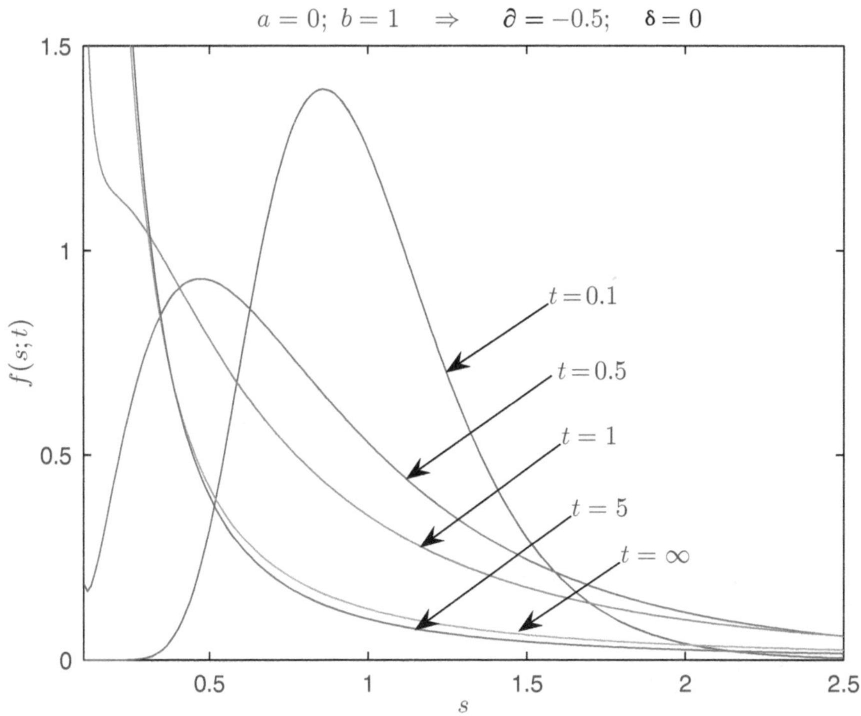

Fig. 7.9 Plots of the pdf $f(s;t)$ given by (7.92) and (7.94) as a function of s for different times, illustrating its convergence to the steady-state pdf $f(s)$ given by (7.87). Here, $s_0 = 1, s_1 = 0.1$, $a = 0$, $b = 1$, so that $\lambda = -1$ satisfies the condition $\lambda < 0$ for the existence of a non-zero steady-state $f(s)$

Figures 7.9 and 7.10 show the pdf $f(s;t)$ obtained from (7.92) and (7.94) for two different regimes. Figure 7.9 corresponds to $\lambda < 0$ for which $f(s;t)$ converges at long times to the non-vanishing steady-state pdf (7.87). Figure 7.10 shows the decay of $f(s;t)$ to zero at long times, for $\lambda > 0$.

Let us now prove that the steady-state density

$$g(s) = \nu \int_0^\infty e^{-\mu u} f(s;u)du\,, \qquad \mu > 0\,, \tag{7.95}$$

exists for any λ. We see from (7.94) that, for $z \neq 0$ (and for $z > z_1$), the pdf $\varphi(z;t)$ is bounded from above while, for $z = 0$, it exhibits an integrable singularity, analogous to (3.21). Thus, there exists, for any $z > z_1$, a well-defined function $\varphi(z)$, which is equal to the convergent improper integral

$$\varphi(z) = \nu \int_0^\infty e^{-\mu u} \varphi(z;u)du\,. \tag{7.96}$$

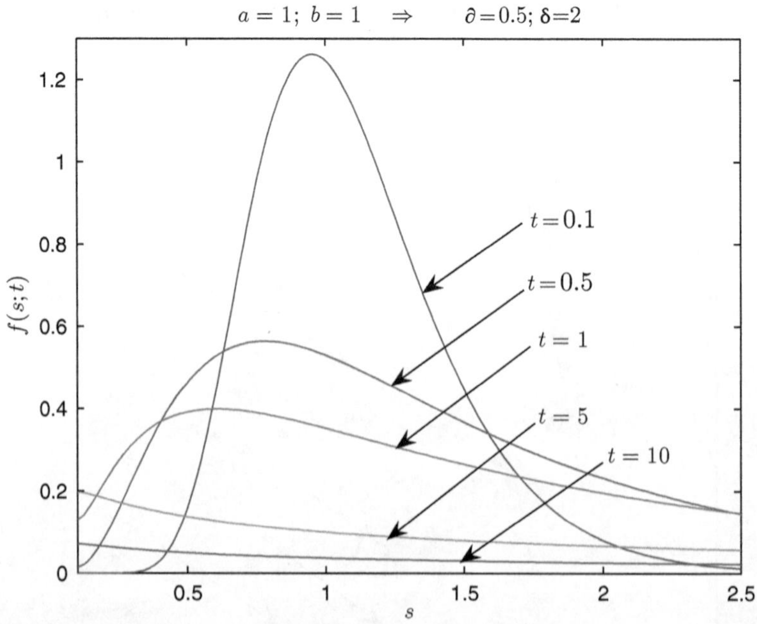

Fig. 7.10 Plots of the pdf $f(s;t)$ given by (7.92) and (7.94), illustrating its uniform convergence to zero, as $t \to \infty$. Here, $s_0 = 1, s_1 = 0.1, a = 1, b = 1$, so that $\lambda = 1$ does not satisfy the condition $\lambda < 0$ for the existence of a non-zero steady-state pdf $f(s)$

In turn, due to relation (7.92), there exists a steady-state density

$$g(s) = \frac{1}{bs}\varphi\left[\frac{1}{b}\ln\left(\frac{s}{s_0}\right)\right] . \tag{7.97}$$

After substitution of (7.94) into (7.96), we obtain[2]

$$\varphi(z) = \frac{1}{\sqrt{2\mu + \varrho^2}} e^{z(\varrho - \sqrt{2\mu+\varrho^2})} \left(1 + e^{2a\sqrt{2\mu+\varrho^2}} \frac{\sqrt{2\mu+\varrho^2} - \varrho}{\sqrt{2\mu+\varrho^2} + \varrho}\right) . \tag{7.98}$$

Then, substituting the last expression into (7.97), we finally get

$$g(s) = \frac{2\nu}{b^2 s_0} A(\lambda, \zeta) \left(\frac{s}{s_0}\right)^{-1-m} \sim s^{-1-m} , \tag{7.99}$$

where

$$A(\delta, \zeta) = \frac{1}{\rho}\left[1 + \left(\frac{s_1}{s_0}\right)^{\rho} \frac{\rho - \lambda}{\rho + \lambda}\right] , \tag{7.100}$$

[2] Related integrals are tabulated in Bateman and Erdelyi (1954).

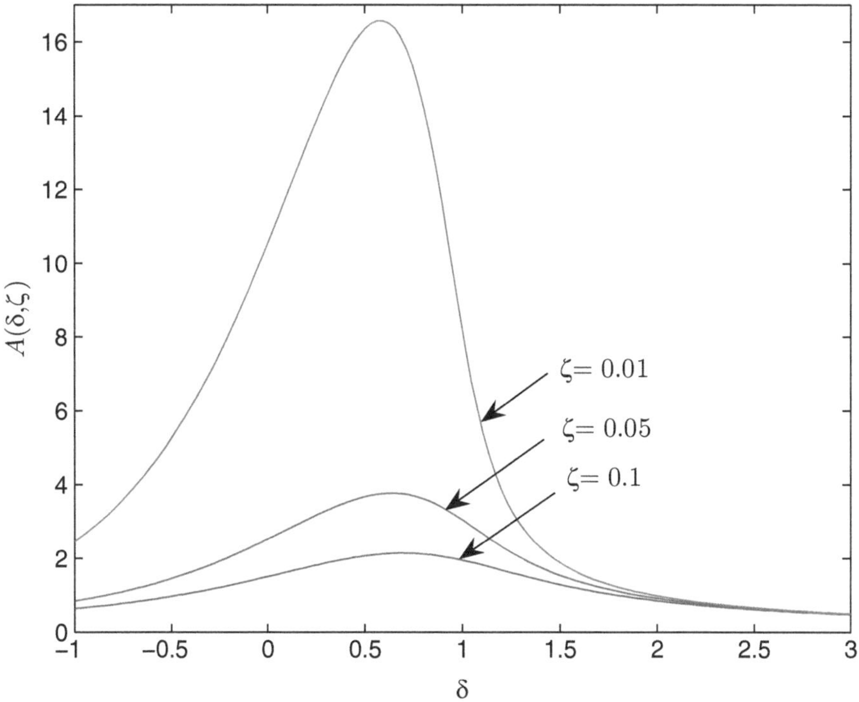

Fig. 7.11 Dependence of the factor $A(\lambda, \zeta)$ defined in (7.100) as a function of λ, for values of parameter $\zeta = 0.1, 0.05, 0.01$, and for $s_1 = 0.1 s_0$

and m is the same as in (7.13)–(7.16):

$$m = m(\lambda, \zeta) = \frac{1}{2}(\rho - \lambda) , \qquad \rho = \sqrt{\lambda^2 + 4\zeta} , \qquad \zeta = \frac{2\mu}{b^2} . \qquad (7.101)$$

Let us compare the steady-state density $g(s)$ given by (7.99) and the steady-state pdf $f(s)$ given by (7.87). Recall that a non-vanishing steady-state pdf $f(s)$ (7.87) exists only if $\lambda < 0$, while a steady-state density $g(s)$ exists for any λ under the condition $\mu > 0$. On the other hand, the aforementioned mutually exclusive existence of $g(s)$ and $f(s)$ for $\mu = 0$ implies that, as $\mu \to 0_+$, the steady-state mean density $g(s)$ (7.99) tends to infinity, if $\lambda < 0$, and remains finite, if $\lambda > 0$.

This warrants studying the limit $\mu \to 0_+$ in more detail. The meaning of μ going to zero can be better understood when interpreted in terms of the expected lifetime μ^{-1} of a typical firm. This lifetime μ^{-1} should be compared with the *volatility time* t_b defined in (7.101), which is the characteristic time of change of the asset's value $S(t)$ due to volatility. The limit $\mu \to 0_+$ corresponds to the condition

$$t_b \ll \mu^{-1} \qquad \Longleftrightarrow \qquad \zeta \ll 1 . \qquad (7.102)$$

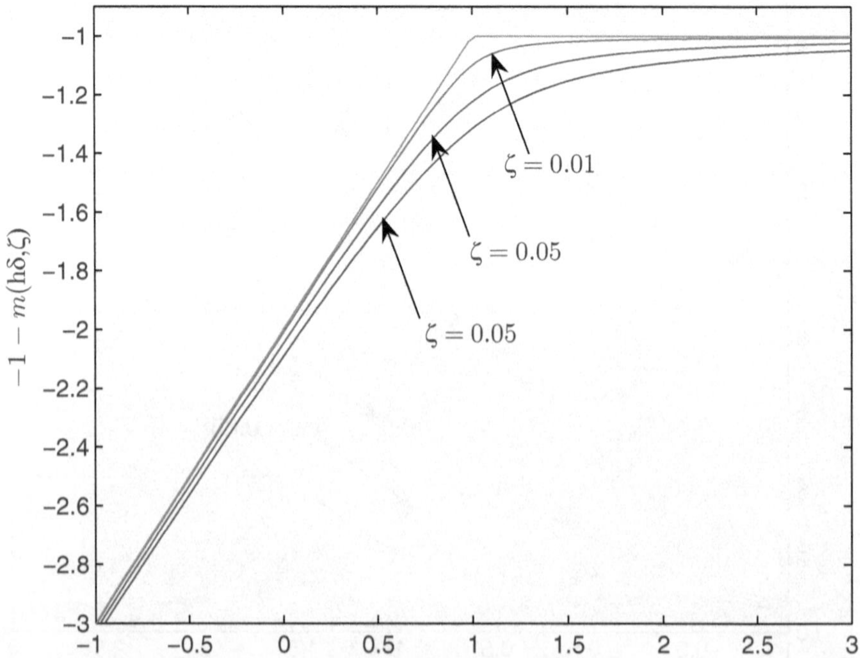

Fig. 7.12 Plots of the tail index $-1-m$ [where m is given in (7.101)] of the steady-state density $g(s)$ (7.99) as a function of λ, for values of the parameter $\zeta = 0.1; 0.05; 0.01$. The *straight line* (for $\lambda < 0$) plots the exponent $\lambda - 1$ of the steady-state pdf $f(s)$ (7.87)

Inequality (7.102) expresses that the characteristic lifetime of a typical firm is much larger than the volatility time t_b over which significant firm value variations occur according to GBM processes. Figure 7.11 shows the dependence of the factor $A(\lambda,\zeta)$ given by (7.100) as a function of λ, for three decreasing values of the parameter ζ, illustrating the announced divergence of $g(s)$ as $\mu \to 0_+$.

Another important comparison is that the steady-state pdf $f(s)$ (7.87) and density $g(s)$ (7.99) are power laws with generally different exponents. But, in the limit where (7.102) holds, the exponents of the two power laws (7.87) and (7.99) become equal as $\zeta \to 0$, in the domain $\lambda < 0$ for which the steady-state pdf $f(s)$ exists. Figure 7.12 plots the exponent $-1-m(\lambda,\zeta)$ (where m is given by (7.101)) of the power law $g(s)$ (7.99) as a function of λ, for different values of ζ, demonstrating that, for $\lambda < 0$ and in the limit $\zeta \to 0$, the power laws (7.87) and (7.99) are actually the same.

Chapter 8
Non-stationary Mean Birth Rate

In all previous chapters, we have studied the steady-state mean density $g(s)$ given by (3.18) of firm's asset values and its properties, for a stationary intensity ν of firm's births. In real life, ν is not constant, with periods of strong growth, such as during "new economy bubbles" (Galbraith, 1997; Kindleberger, 2000; Shefrin, 2000; Shiller, 2000; Shleifer, 2000; Sornette, 2003; White, 1996) or during and after political transitions, and periods of stagnation, for instance during depressions. Over large times, there are even secular variations of firms creations, such as for instance during the transition associated with the political "big bang" of the Soviet Bloc in the 1990s (Nowak et al., 2005). In some countries, (e.g., Poland), not long after the transition, the economy started to grow at a fast rate soon surpassing the level of its economy under socialism, with a large growth of the number of privately owned enterprises during the transition from centrally governed to the market economy (Nowak et al., 2000, 2005; Gur et al., 2008).

In this chapter, we first derive some properties resulting from a non-stationary birth intensity $\nu(t)$ of the mean density $g(s,t)$ of firm's asset values given by (3.15). Then, we introduce and study a model in which the intensity of firm's birth is coupled with the overall firm's asset value: as the later grows, the former is also assumed to grow correspondingly. This simple model accounts more realistically for the fact that firm's creation is indeed related to the innovation dynamics and capital availability, both being stronger in periods of firm's growth.

8.1 Exponential Growth of Firm's Birth Rate

Let us assume that the intensity of firm's births is exponentially growing as

$$\nu(t) = \nu_0 e^{dt}, \qquad d > 0. \tag{8.1}$$

Substituting this expression into (3.16), we obtain

$$g(s,t) = e^{dt} \int_0^t e^{-du} f(s;u)du = \nu(t) \int_0^t e^{-du} f(s;u)du. \tag{8.2}$$

A. Saichev et al., *Theory of Zipf's Law and Beyond*,
Lecture Notes in Economics and Mathematical Systems 632,
DOI 10.1007/978-3-642-02946-2_8, © Springer-Verlag Berlin Heidelberg 2010

This formula (8.2) differs from expression (7.6) only by the replacement of ν by $\nu(t)$. Thus, all the properties of the mean density of firm's asset values which were previously derived for a constant ν, including those taking into account the possibility of sudden firm's death, carry over to the present case with a $\nu(t)$ varying exponentially with time. In particular, the generalized balance condition stated in Proposition 7.4.1 (with replacing μ by d) holds true, although it acquires another interpretation.

Consider firms which were born at some instant $t_0 = t - u$, so that at the current time t their mean asset values are equal to $E[S(u)]$. Correspondingly, the overall mean asset value $\Omega(t, u)$, at the current instant t, of all firms which were born at instant $t - u$, is proportional to

$$\Omega(t, u) \sim \nu(t - u)\, E[S(u)]. \tag{8.3}$$

Supposing for simplicity that $S(t)$ is a GBM whose expectation is given by equality (7.29), and taking into account relation (8.1), we obtain

$$\Omega(t, u) \sim e^{d(t-u)} s_0 e^{au} = s_0 e^{dt}\, e^{(a-d)u}. \tag{8.4}$$

If condition $a = d$, which is formally equivalent to the generalized balance condition (Proposition 7.4.1), is valid, then the overall mean asset value at current time t

$$\Omega(t, u) \sim s_0 e^{dt}, \tag{8.5}$$

does not depend on u. In other words, equality $a = d$ is a balance condition in the sense that a group of firms which were born at the time t_1 ($t_1 < t$) has the same mean overall asset value as another groups of firms which were born at any other instant t_2 ($t_2 < t$).

8.2 Deterministic Skeleton of Zipf's Law

We can now reveal more transparently the hidden working of Zipf's law for the mean density of firm's asset values, where the generalized balance condition is valid, based on the deterministic version of the distribution of firm's asset values. Let $S(t - t')$ be a deterministic function describing the growth of the asset value of some firm which was born at instant t', while $\nu(t)$ is the intensity of firm's birth. Then the overall number of firms, whose asset values at current time t are larger than some level s, is equal to

$$G(s; t) = \int_0^t \nu(t')\mathbf{1}\left[S(t - t') - s\right] dt', \tag{8.6}$$

where $\mathbf{1}(x)$ is the unit step function. Let $S(t)$ be an increasing function, so that the integral (8.6) is equal to

$$G(s;t) = \int_0^{t-t^*(s)} \nu(t')\,dt', \tag{8.7}$$

where $t^*(s)$ is the age of the firm upon reaching asset value s. One can find t^* by solving the equation

$$S(t^*) = s \tag{8.8}$$

with respect to t^*. In particular, if the deterministic version of Gibrat's law is true, then we have

$$S(t) = s_0 e^{at} \qquad \Longrightarrow \qquad t^*(s) = \frac{1}{a} \ln\left(\frac{s}{s_0}\right), \qquad a > 0. \tag{8.9}$$

From another point of view, if the intensity of firm's birth grows exponentially, i.e., if it is given by expression (8.1), then it follows from (8.7) that

$$G(s;t) = \frac{\nu(t)}{d} \cdot e^{-dt^*(s)}. \tag{8.10}$$

Expression (8.10) implies that, the larger the exponent d of firm's births, and the greater the age $t^*(s)$ of firms upon reaching the level s, the smaller the fraction of firms whose asset values exceed s.

One may interpret the exponential growth (8.1) as some kind of Gibrat's law for the intensity of firm's birth. Substituting into (8.10) the consequence of a deterministic Gibrat's law for firm's asset values given by (8.9), we obtain the power law

$$G(s;t) = \frac{\nu(t)}{d}\left(\frac{s_0}{s}\right)^{\varepsilon}, \qquad \varepsilon = \frac{d}{a}. \tag{8.11}$$

If both Gibrat's laws are consistent, in the sense that their growth rates are identical ($a = d$), then we get Zipf's law.

8.3 Simple Model of Birth Rate Coupled with the Overall Firm's Value

The properties of the mean density of firm's asset values, when Gibrat's law holds exactly or approximately in some asymptotic regimes, have been studied by considering that the three controlling processes are decoupled and can be chosen independently to represent different properties of the economics of firm's dynamics. These three processes are:

1. The instantaneous growth rate $a(t)$ of the firm's asset values
2. The volatility $b(t)$
3. The intensity of firm's birth $\nu(t)$

However, as discussed in the introduction of this chapter, it seems reasonable to consider the possibility that some or all three processes are coupled. For instance, as the overall value of the firms belonging to an economy grows, the availability for innovations and capital to seed new firms should also grow concomitantly.

In the sequel, we introduce the simplest model that takes into account a possible coupling between the birth rate of firms at a given time and the distribution of firm's values as a function of time, which is taken to characterize the evolution with time of the state of the economy. The model is defined by

$$\nu(t) = \nu_0(t) + \varsigma\,\mathcal{K}(t)\,, \qquad \mathcal{K}(t) := \int_0^\infty \theta(s) g(s,t) ds. \tag{8.12}$$

Here, ς is a coupling factor, while the term $\nu_0(t)$ describes the spontaneous appearance of some "pioneering firms," arising due to new inventions, new technologies, and new niches, giving rise to a new area of economic development. In what follows, we will suppose that

$$\nu_0(t) = 0 \qquad \text{for all} \qquad t < t_0, \tag{8.13}$$

where t_0 is the time of foundation of the new economic era. Correspondingly, we will suppose, in accordance with the causality principle, that both the intensity of firm's birth and the mean density of firm's asset values are equal to zero earlier than the foundation time:

$$\nu(t) = 0\,, \qquad g(s,t) = 0\,, \qquad \text{for all} \qquad t < t_0. \tag{8.14}$$

The function $\theta(s)$ in (8.12) describes the mechanism(s) for the initialization of new firms that result from the coupling with the current existence of firms. For instance, in a situation in which the greater the number of firms which possess assets whose values are larger than some given value s_*, the larger the probability of firm's births, then $\theta(s) = \mathbf{1}(s - s_*)$ and relation (8.12) becomes

$$\nu(t) = \nu_0(t) + \varsigma \mathcal{N}(s_*, t) = \nu_0(t) + \varsigma \int_{s_*}^\infty g(s,t) ds, \tag{8.15}$$

where $\mathcal{N}(s_*, t)$ is the average number of firms whose asset values at instant t are larger than s_*. In a different economy in which the probability of the birth of firms increases with the overall firm asset value, then $\theta(s) = s$, which leads to the following expression

$$\nu(t) = \nu_0(t) + \varsigma \Omega(t) = \nu_0(t) + \varsigma \int_{s_1}^\infty s g(s,t) ds, \tag{8.16}$$

where s_1 is the lowest possible asset value.

We have seen that the mean density $g(s,t)$ of firm's asset values is given by Corollary 3.2.1, which has to be complemented with the conditions (8.14). We can thus write

$$g(s,t) = \int_{t_0}^{t} \nu(t') f(s; t-t') dt'. \tag{8.17}$$

Equation (8.16) together with (8.17) constitute two coupled integral equations in the functions $\nu(t)$ and $g(s,t)$.

We first consider the general case. Substituting (8.17) into (8.12), we obtain the following integral equation with respect to the mean birth rate $\nu(t)$:

$$\nu(t) = \nu_0(t) + \varsigma\, \mathcal{R}(t) \otimes \nu(t), \tag{8.18}$$

where

$$\mathcal{R}(t) := \int_0^{\infty} \theta(s) f(s; t) ds. \tag{8.19}$$

Without loss of generality, we have put $t_0 = 0$ in (8.18), and have used the sign $\otimes$ for the convolution integral.

Using the Laplace transform

$$\hat{\nu}(u) := \int_0^{\infty} \nu(t) e^{-ut} dt \tag{8.20}$$

of $\nu(t)$ allows us to transform the integral equation (8.18) for the unknown intensity of firm's birth $\nu(t)$ into an algebraic equation for its Laplace transform $\hat{\nu}(u)$:

$$\hat{\nu}(u) = \hat{\nu}_0(u) + \varsigma\, \hat{\mathcal{R}}(u) \cdot \hat{\nu}(u). \tag{8.21}$$

Thus, the explicit expression for the Laplace transform of the intensity of firm's birth is

$$\hat{\nu}(u) = \frac{\hat{\nu}_0(u)}{1 - \varsigma \hat{\mathcal{R}}(u)}. \tag{8.22}$$

We can now explore some properties of the intensity of firm's birth $\nu(t)$ for a case appropriate for many applications, such that $\mathcal{R}(t)$ is proportional to the firm's mean asset value

$$\mathcal{R}(t) = \int_{s_1}^{\infty} s f(s; t) ds = \mathrm{E}\left[S(t)\right]. \tag{8.23}$$

As already mentioned, this situation describes an economy in which $\theta(s) = s$. Taking $s_1 = 0$ for simplicity, and supposing that the stochastic behavior of the asset value $S(t)$ of any firm obeys the pure Gibrat's law, we obtain

$$\mathcal{R}(t) = s_0 e^{at} \qquad \Longrightarrow \qquad \hat{\mathcal{R}}(u) = \frac{s_0}{u-a}. \tag{8.24}$$

Let us assume that the birth intensity of pioneering firms $\nu_0(t)$ is given by

$$\nu_0(t) := \nu_0 e^{-dt} \mathbf{1}(t) \qquad \Longrightarrow \qquad \hat{\nu}_0(u) = \frac{\nu_0}{u+d}, \tag{8.25}$$

corresponding to a burst of innovation at the origin of the new technological era, followed by a progressive exponential relaxation. Substituting the Laplace images (8.24) and (8.25) into (8.22), we obtain

$$\hat{\nu}(u) = \nu_0 \frac{u-a}{(u+d)(u-a-\varsigma s_0)}. \tag{8.26}$$

This leads to

$$\nu(t) = \nu_0 \frac{1+\lambda+\rho}{1+\lambda+\rho+\gamma} e^{-\rho\tau} + \nu_0 \frac{\gamma}{1+\lambda+\rho+\gamma} e^{(1+\lambda+\gamma)\tau}, \tag{8.27}$$

where we have used the following dimensionless parameters

$$\gamma := \frac{2\varsigma s_0}{b^2}, \qquad \rho := \frac{2d}{b^2}. \tag{8.28}$$

The first term in the r.h.s. of (8.27) describes the impact of pioneering firms, whose rate of birth is obtained by a renormalization of the bare exponent d by the stochastic diffusive terms of the GBM dynamics of firm's asset values. The second term in the r.h.s. of (8.27) results from the cascades of firms born as the economy develops, and reveals the coupling between the multiplicative Gibrat growth and the feedback of the growth of the population of firms on the creation of new firms.

Since the first term in the r.h.s. of (8.27) decays exponentially, we can omit it at long times ($\tau \gg 1/\rho$), leading to the asymptotic expression

$$\nu(t) \simeq N_0 \, e^{(1+\lambda+\gamma)\tau} \qquad (1+\lambda > -\gamma), \qquad N_0 := \nu_0 \frac{\gamma}{1+\delta+\rho+\gamma}. \tag{8.29}$$

Expression (8.29) describes the intensity of firm's birth in a "developed market." Correspondingly, in accordance with (7.15), the mean density of firm's asset values in a developed market regime obeys the asymptotic power law

$$g(s) \sim s^{-(1+m)}, \qquad s > s_0, \tag{8.30}$$

where the exponent takes the new expression

$$m = \frac{1}{2}\left(\sqrt{(2+\lambda)^2 + 4\gamma} - \lambda\right). \tag{8.31}$$

Zipf's law is recovered in the limit $\gamma \to 0$. Again, we see that Zipf's law is a robust outcome of firm's growth characterized by large stochasticity (b large relative to $\sqrt{2\varsigma s_0}$). For small γ's, this economy is thus described by an approximate Zipf's law

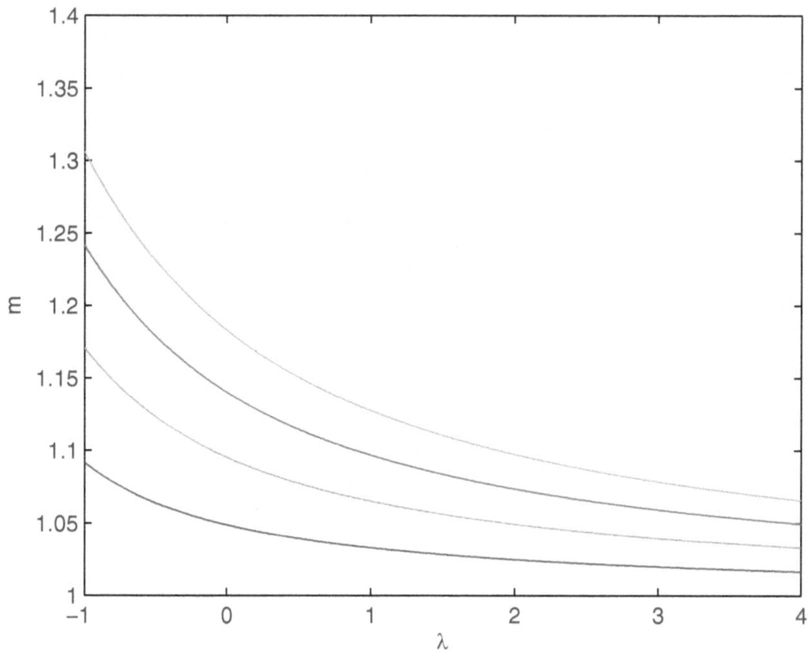

Fig. 8.1 Plot of the dependence of the exponent m given by (8.31) as a function of λ for different values of the dimensionless parameter γ. *From bottom to top*, $\gamma = 0.1, 0.2, 0.3, 0.4$

($m \approx 1$) for arbitrary values of λ. Figure 8.1 shows the dependence of m given by (8.31) as a function of λ, for different values of the parameter γ.

8.4 Generalization When Both the Initial Firm's Sizes and the Minimum Firm's Size Grow at Constant Rates

8.4.1 Formulation of the Model

In order to generalize the results derived in the previous sections, and to be closer to the real world, we now assume that, in addition to the intensity of firm's births, both the initial size of an entrant firm and the minimum size of an incumbent firm grow at the constant rates c_0 and c_1 respectively. As a consequence, assuming without loss of generality that the economy starts at $t = 0$, the initial size of an entrant firm at time t is given by

$$s_0(t) := s_0 \, e^{c_0 t} . \tag{8.32}$$

Similarly, the minimum size of a firm at time t, below which the firm exits, is

$$s_1(t) := s_1 \, e^{c_1 t} , \qquad c_1 \le c_0 , \qquad s_1 \le s_0 . \tag{8.33}$$

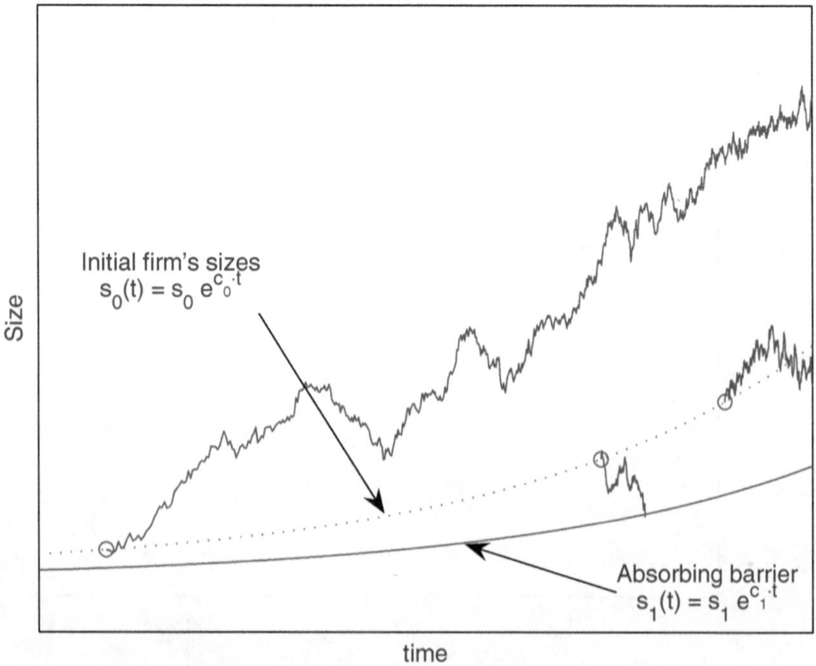

Fig. 8.2 Three firm birth events (followed by the size dynamics of the three firms) are shown at three different times, to illustrate the effect of the constant growth rate c_0 of initial firm sizes. A death event is also represented and illustrates the existence of the constant growth rate c_1 of the minimum firm size.

In other words, when at some instant t_e, the firm size "touches" for the first time the exit level $s_1(t)$, the firm dies. This generalization of the model will allow us to better understand the underpinning of the balance condition which has appeared up to now as the keystone constraining the tail index of the firm size distribution to the value 1 (Zipf's law). Figure 8.2 illustrates the model.

Let us consider a firm born at time u, whose size at the current time t, denoted by $S(t,u)$, is given by the following stochastic process

$$S(t,u) = s_0(u)e^{c(t-u)+bW(t-u)}, \tag{8.34}$$

where $W(t)$ is a standard Wiener process, while $s_0(u)$ is the initial size of the firm.

The process (8.34) with the conditions (8.32) and (8.33) can be reformulated as

$$S(t,u) = s_1(t)e^{Z(t,u)}, \tag{8.35}$$

where

$$Z(t,u) := z_*(u) + (c - c_1)(t-u) + bW(t-u), \tag{8.36}$$

is an auxiliary Wiener process with drift, and

$$z_*(u) := \ln\left(\frac{s_0}{s_1}\right) + (c_0 - c_1)u. \tag{8.37}$$

The first condition (8.32) on the initial size of firms is accounted for in the auxiliary Wiener process $Z(t,u)$ with drift. The firm's death occurring when $S(t,u)$ touches for the first time the value $s_1(t)$ given by (8.33) is now replaced by the condition that $Z(t,u)$ touches the zero level $Z = 0$. We derive below the statistical properties of the stochastic firm's size $S(t,u)$ from the known statistical properties of the stochastic process $Z(t,u)$.

Both the auxiliary stochastic process $Z(t,u)$ and the original stochastic firm's size $S(t,u)$ depend on the current time t and on the birthdate u. For our following calculations, it is more convenient to express these processes in terms of t and of the current age of the firm

$$\theta := t - u > 0. \tag{8.38}$$

We then have

$$Z(t,u) = \mathcal{Z}(t,\theta) \qquad \Longrightarrow \qquad S(t,u) = s_1(t)\, e^{\mathcal{Z}(t,\theta)}, \tag{8.39}$$

where

$$\mathcal{Z}(t,\theta) := z_*(u) + (c - c_1)\theta + bW(\theta). \tag{8.40}$$

Since $z_*(u) = \ln \rho(t) + (c_1 - c_0)\theta$, with

$$\rho(t) := \frac{s_0(t)}{s_1(t)}, \tag{8.41}$$

we obtain

$$\mathcal{Z}(t,\theta) = \ln \rho(t) + (c - c_0)\theta + bW(\theta). \tag{8.42}$$

The change of variables $(t,u) \to (t,\theta)$ allows us to analyze the statistical properties of firm's sizes at a fixed time t as a function of their ages θ. Denoting as above by t_e the time at which the firm disappears due to $\mathcal{Z}(t,\theta)$ touching the zero level, we need to take into account only those realizations of the stochastic process $\mathcal{Z}(t,\theta)$, for which $t < t_e$. Equivalently, this amounts to studying only those realizations of the stochastic process $\mathcal{Z}(t,\theta)$ (8.42) such that $\theta < \theta_e$, where $\theta_e = t_e - u$ is the age of the firm at death.

For the model to be consistent, the initial firm size must be always larger than the minimum firm size. This implies that the economy started at a time not earlier than u_* given by

$$u_* := \frac{1}{c_1 - c_0} \ln\left(\frac{s_0}{s_1}\right), \qquad u_* > -\infty. \tag{8.43}$$

Therefore, the oldest firm was born at u_*. At time t, its age is

$$\theta_*(t) = \frac{\ln \rho(t)}{c_0 - c_1}, \tag{8.44}$$

where $\rho(t)$ is defined in (8.41).

8.4.2 Pdf $f(s;t,\theta)$ of Firm's Size

Let us now determine the pdf $f(s;t,\theta)$ of the firm sizes at age θ, taking into account a possible firm's death when its size reaches the exit level $s_1(t)$. It follows from (8.39) that $f(s;t,\theta)$ is obtained from the pdf $\varphi(z;t,\theta)$ of the stochastic process $\mathcal{Z}(t,\theta)$ (8.40) through the relationship

$$f(s;t,\theta) = \frac{1}{s}\varphi\left[\ln\left(\frac{s}{s_1(t)}\right);t,\theta\right]. \tag{8.45}$$

We thus need to derive the pdf $\varphi(z;t,\theta)$ of the stochastic process $\mathcal{Z}(t,\theta)$ (8.42). Interpreting $\mathcal{Z}(t,\theta)$ as a Wiener process with drift with respect to the variable θ, $\varphi(z;t,\theta)$ is found as the solution to the initial-boundary problem

$$\begin{gathered} \frac{\partial\varphi(z;t,\theta)}{\partial\theta} + (c-c_1)\frac{\partial\varphi(z;t,\theta)}{\partial z} = \frac{b^2}{2}\frac{\partial^2\varphi(z;t,\theta)}{\partial z^2}, \\ \varphi(z;t,\theta=0) = \delta(z - \ln\rho(u)), \\ \varphi(z=0;t,\theta) = 0\,, \qquad \theta > 0. \end{gathered} \tag{8.46}$$

Using the reflection method, the solution to (8.46) is

$$\begin{aligned} \varphi(z;t,\theta) =& \frac{1}{\sqrt{2\pi b^2\theta}}\exp\left(-\frac{(z-\ln\rho(t)-(c-c_1)\theta)^2}{2b^2\theta}\right) \\ &- \frac{A}{\sqrt{2\pi b^2\theta}}\exp\left(-\frac{(z+\ln\rho(t)-(c-c_1)\theta)^2}{2b^2\theta}\right), \end{aligned} \tag{8.47}$$

where the constant A is determined so that $\varphi(z;t,\theta)$ satisfies the boundary condition $\varphi(z=0;t,\theta) = 0$ defined in (8.46):

$$A := \rho^{\lambda_1-\lambda}(t), \tag{8.48}$$

where

$$\lambda_0 := \frac{2c_0}{b^2}\,, \quad \lambda_1 := \frac{2c_1}{b^2}\,, \quad \tau := \frac{b^2}{2}\theta\,, \quad \tau_* := \frac{b^2}{2}\theta_*(t). \tag{8.49}$$

This leads finally to

$$\varphi(z;t,\theta) = \frac{1}{2\sqrt{\pi\tau}}\left[\exp\left(-\frac{(z - \ln\rho(t) - (\lambda-\lambda_1)\tau)^2}{4\tau}\right)\right.$$
$$\left. - \rho(t)^{\lambda_1-\lambda}\exp\left(-\frac{(z + \ln\rho(t) - (\lambda-\lambda_1)\tau)^2}{4\tau}\right)\right]. \quad (8.50)$$

Below, we restrict ourself by discussing only the most natural case $c_0 = c_1$, for instance to account for a global impact of inflation leading to the same growth rate for the initial firm sizes $s_0(t)$ and the exit boundary $s_1(t)$.

In this case, one can rewrite the previous expression in the form

$$\varphi(z;t,\theta) = \frac{1}{2\sqrt{\pi\tau}}\left[\exp\left(-\frac{(z_- - \alpha\tau)^2}{4\tau}\right) - \rho(t)^{-\alpha}\exp\left(-\frac{(z_+ - \alpha\tau)^2}{4\tau}\right)\right], \quad (8.51)$$

where

$$z_\pm = z \pm \ln\rho(t) \qquad \text{and} \qquad \alpha = \lambda - \lambda_0. \quad (8.52)$$

8.4.3 Mean Density $g(s,t)$ of Firm Sizes

The mean density of firm sizes at the current time t is equal to the following integral over all possible firm's birthdates u

$$g(s,t) = \int_{u_0}^{t} \nu(u)\mathcal{Q}(t-u)f(s;t,t-u)du\,, \qquad t > t_0, \quad (8.53)$$

where t_0 is the birthdate of the given economy, $\nu(t)$ is the intensity of firm's births, and $\mathcal{Q}(\theta)$ is the survival function, i.e., the probability for a firm to survive till the age θ. As in Chap. 7, we assume that both $\nu(t)$ and $\mathcal{Q}(\theta)$ are exponential:

$$\nu(t) = \nu\, e^{dt}\,, \quad \text{and} \quad \mathcal{Q}(\theta) = e^{-\mu t}\,, \qquad \mu > 0, \quad (8.54)$$

where μ is the hazard rate. Using these expressions in (8.53) and performing the change of variable from birthdate u to age $\theta = t - u$ leads to

$$g(s,t) = \nu(t)\int_0^{\theta_0} e^{-(d+h)\theta} f(s;t,\theta)d\theta, \quad (8.55)$$

where $\theta_0 = t - u_0$ is the age of the given economy. Inasmuch as u_0 should not be smaller than u_* given by (8.43), we should thus have $\theta_0 < \theta_*$.

Substituting (8.45) into (8.55), we obtain

$$g(s,t) = \tilde{\nu}(t)\,\frac{1}{s}\,\tilde{g}\left(\ln\left(\frac{s}{s_1(t)}\right); t, \tau_*, \tau_0\right), \qquad \tilde{\nu}(t) := \frac{2\nu(t)}{b^2}, \quad (8.56)$$

where

$$\tilde{g}(z;t,\tau_*,\tau_0) := \int_0^{\tau_0} e^{-\eta\tau}\varphi(z;t,\theta)d\tau, \tag{8.57}$$

with

$$\tau_0 := \frac{b^2}{2}\,\theta_0 \qquad (\tau_0 < \tau_*)\,, \qquad \eta := \frac{2}{b^2}\,(d+\mu). \tag{8.58}$$

The substitution of $\varphi(z;t,\theta)$ (8.51) into the integral (8.57) leads to two integrals, which can be reduced to the following integral function

$$\mathcal{I}(z,\theta,\alpha,\beta) := \int_0^{\theta} \exp\left(-\frac{(z-\alpha\tau)^2}{4\tau} - \beta\tau\right)\frac{d\tau}{2\sqrt{\pi\tau}}. \tag{8.59}$$

This integral function $\mathcal{I}(z,\theta,\alpha,\beta)$ generalizes the integral function

$$\mathcal{I}(z,\alpha,\beta) = \mathcal{I}(z,\theta=\infty,\alpha,\beta) = \int_0^{\infty} \exp\left(-\frac{(z-\alpha\tau)^2}{4\tau} - \beta\tau\right)\frac{d\tau}{2\sqrt{\pi\tau}}. \tag{8.60}$$

Performing the integration in (8.60) leads to

$$\mathcal{I}(z,\alpha,\beta) = \frac{1}{\sqrt{\alpha^2+4\beta}}\exp\left[\frac{1}{2}\left(\alpha z - \sqrt{\alpha^2+4\beta}\,|z|\right)\right], \qquad (\alpha^2+4\beta>0). \tag{8.61}$$

Using the tabulated integral (7.4.33) in Abramowitz and Stegun (1964) yields

$$\begin{aligned}\mathcal{I}(z,\theta,\alpha,\beta) =& \frac{1}{2\,\alpha(\beta)}\left[\exp\left[\frac{1}{2}\left(\alpha z - \alpha(\beta)\,|z|\right)\right]\operatorname{erfc}\left(\frac{|z|-\theta\alpha(\beta)}{2\sqrt{\theta}}\right)\right.\\ &\left. - \exp\left[\frac{1}{2}\left(\alpha z + \alpha(\beta)\,|z|\right)\right]\operatorname{erfc}\left(\frac{|z|+\theta\alpha(\beta)}{2\sqrt{\theta}}\right)\right],\end{aligned} \tag{8.62}$$

where

$$\alpha(\beta) := \sqrt{\alpha^2+4\beta}\,. \tag{8.63}$$

Formula (8.62) holds true even for imaginary $\alpha(\beta)$ (obtained for $4\beta < -\alpha^2$). We verify directly that expression (8.62) recovers (8.61) by using the following asymptotic relations

$$\begin{aligned}\operatorname{erfc}(x) &\simeq e^{-x^2}\left(\frac{1}{\sqrt{\pi}x} + O\left(\frac{1}{x^3}\right)\right), \qquad x\to\infty,\\ \operatorname{erfc}(x) &\simeq 2 + e^{-x^2}\left(\frac{1}{\sqrt{\pi}x} + O\left(\frac{1}{x^3}\right)\right), \qquad x\to-\infty.\end{aligned} \tag{8.64}$$

Substituting in expression (8.57) for $\tilde{g}(z;t,\tau_*,\tau_0)$ the function $\varphi(z;t,\theta)$ given by (8.51) and using the analytical form (8.62) for the integral function (8.59), we obtain

$$\begin{aligned}\tilde{g}(z;t,\tau_0) =& \frac{1}{2\alpha(\eta)}\Bigg\{ e^{\frac{1}{2}(\alpha z_- - \alpha(\eta)|z_-|)}\text{erfc}\left(\frac{|z_-| - \tau_0\alpha(\eta)}{2\sqrt{\tau_0}}\right) \\ & - e^{\frac{1}{2}(\alpha z_- + \alpha(\eta)|z_-|)}\text{erfc}\left(\frac{|z_-| + \tau_0\alpha(\eta)}{2\sqrt{\tau_0}}\right) - \rho(t)^{-\alpha} \\ & \times \left[e^{\frac{1}{2}(\alpha z_+ - \alpha(\eta)|z_+|)}\text{erfc}\left(\frac{|z_+| - \tau_0\alpha(\eta)}{2\sqrt{\tau_0}}\right)\right. \\ & \left.\left. - e^{\frac{1}{2}(\alpha z_+ + \alpha(\eta)|z_+|)}\text{erfc}\left(\frac{|z_+| + \tau_0\alpha(\eta)}{2\sqrt{\tau_0}}\right)\right]\right\}. \end{aligned} \tag{8.65}$$

In the case where $\alpha(\eta)\sqrt{\tau_0} \gg 1$, we can expand the above expression for $G(z;t,\tau_0)$ using the asymptotic (8.64) and obtain

$$\tilde{g}(z;t,\tau_0) = \frac{1}{\alpha(\eta)}\left[e^{\frac{1}{2}(\alpha z_- - \alpha(\eta)|z_-|)} - \rho(t)^{-\alpha}e^{\frac{1}{2}(\alpha z_+ - \alpha(\eta)|z_+|)}\right]. \tag{8.66}$$

We now substitute this last expression into (8.56) for the mean density of firm's sizes, after making explicit the s-dependence of the variable z according to $z = \ln\left(\frac{s}{s_1(t)}\right)$. Using the following notations summarizing our definitions in (8.41) and (8.52),

$$z_- = \ln\kappa\,, \qquad z_+ = \ln(\kappa\rho^2)\,, \qquad \kappa = \frac{s}{s_0(t)}\,, \qquad \rho = \frac{s_0(t)}{s_1(t)}, \tag{8.67}$$

we thus obtain finally, for large $\tau_0 \gg 1$,

$$g(s,t) = \frac{\tilde{\nu}(t)}{s\alpha(\eta)}\begin{cases}\kappa^{\frac{1}{2}(\alpha-\alpha(\eta))}\left(1-\rho^{-\alpha(\eta)}\right), & \kappa > 1, \\ \kappa^{\frac{1}{2}(\alpha+\alpha(\eta))} - \rho^{-\alpha(\eta)}\kappa^{\frac{1}{2}(\alpha-\alpha(\eta))}, & 1 > \kappa > \rho^{-1}.\end{cases} \tag{8.68}$$

8.4.4 Local Principle

Before discussing the conditions controlling the value of the exponent of the asymptotic power law describing the tail of the mean density $g(s,t)$, we point out important properties of the general expression for $g(s,t)$ given by (8.56) with (8.65), as well as its limiting case (8.68).

The general expression of the mean density given by (8.56) with (8.65) does not depend explicitly on the (dimensionless) maximum age τ_* of the economy, but only on the (dimensionless) time τ_0, corresponding to the lower limit of the integral (8.53) associated with the birthdate u_0 of the oldest firm in the economy. As discussed above, for our results to hold, we must have $\tau_0 \leqslant \tau_*$. It is therefore natural to identify τ_0 with τ_*, and we will drop the index 0 or $*$ in the following.

The mean density $g(s,t)$ determined by (8.56) with (8.65) as a function of the reduced variable $\kappa := s/s_0(t)$ depends on the initial firm sizes $s_0(u)$ and on the exit level $s_1(t)$ only through their combination $\rho(t) = s_0(t)/s_1(t) > 1$ *at the current observation time* t. We refer to this remarkable fact as

Proposition 8.4.1 (Local Principle). *The mean density $g(s,t)$ of firm's sizes at current time t depends only on the ratio $\rho(t)$ of the initial size to the exit level* at the same current time t, *and does not depend on the growth rate c_1 of the exit level.*

A first consequence of the Local Principle is that our results hold for any time dependence of the exit level, as long as $\rho(t)$ remains larger than 1. A second consequence of the local principle is that the time dependence of the mean density $g(s,t)$ is fully captured in the scaling variable κ, together with the ratio ρ at current time, and the (dimensionless) age τ of the economy. Formally, these statements are summarized as follows. The normalized mean density

$$\mathcal{G}(\kappa, \alpha, \eta, \rho(t), \tau) = \frac{g(\kappa s_0(t), t)}{\tilde{\nu}(t) s_0(t)} \tag{8.69}$$

depends only on the following five dimensionless parameters:

1. Normalized firm size $\kappa = \frac{s}{s_0(t)}$.
2. Dimensionless relative drift $\alpha = \lambda - \lambda_0 = \frac{2}{b^2}(c - c_0)$.
3. Composed birth and death hazard rates $\eta = \frac{2}{b^2}\,(d + \mu)$.
4. Current ratio of the initial size and exit level $\rho(t) = \frac{s_0(t)}{s_1(t)}$.
5. Current normalized overall age of the economy $\tau = \frac{b^2}{2}(t - u_*)$, where u_* given by (8.43) is the birthdate of the economy.

8.4.5 Power Law Exponent and Balance Condition

Expression (8.69) with (8.68) shows that the normalized mean density of firm's sizes has the following upper power tail

$$\mathcal{G}(\kappa, \alpha, \eta, \rho, \tau) \sim \kappa^{-m-1}, \qquad \kappa > 1 \qquad (s > s_0(t)), \tag{8.70}$$

where

$$m = \frac{1}{2}\left(\sqrt{\alpha^2 + 4\eta} - \alpha\right). \tag{8.71}$$

Zipf's law, which corresponds to the case $m = 1$, is recovered under the *balance condition*:

$$m = 1 \qquad \Rightarrow \qquad \alpha - \eta = -1, \tag{8.72}$$

which translates in terms of original rates as

$$c - c_0 - \mu - d = -\frac{b^2}{2}. \tag{8.73}$$

Introducing the average growth rate of a firm's size

$$a = c + \frac{b^2}{2}, \tag{8.74}$$

the balance condition (8.73) now reads

$$a = \mu + d + c_0 \qquad \Longleftrightarrow \qquad \mu = a - c_0 - d \tag{8.75}$$

Again, notice that the balance condition does not depend on the growth rate c_1 of the exit level.

In the limiting case $\tau = \infty$, $\mathcal{G}(\kappa, \alpha, \eta, \rho, \tau)$ takes the simple form

$$\mathcal{G}(\kappa, \alpha, \eta, \rho, \tau \to \infty) = \frac{1}{\alpha(\eta)\,\kappa} \begin{cases} \kappa^{-m_-}\left(1 - \rho^{-\alpha(\eta)}\right), & \kappa > 1\,, \\ \kappa^{m_+} - \rho^{-\alpha(\eta)}\kappa^{-m_-}, & 1 > \kappa > \rho^{-1}, \end{cases} \tag{8.76}$$

where

$$m_- := \frac{1}{2}\left(\alpha(\eta) - \alpha\right), \qquad m_+ := \frac{1}{2}\left(\alpha(\eta) + \alpha\right), \qquad \alpha(\eta) = \sqrt{\alpha^2 + 4\eta}. \tag{8.77}$$

When the balance condition (8.72) is valid, then $m_- = 1$, $m_+ = \eta$ and expression (8.76) takes the simple form

$$\mathcal{G}(\kappa, \alpha, \eta, \rho) = \frac{1}{1+\eta} \begin{cases} \kappa^{-2}(1 - \rho^{-\eta-1}), & \kappa > 1 \\ \kappa^{\eta-1} - \rho^{-\eta-1}\kappa^{-2}, & 1 > \kappa > \rho^{-1}. \end{cases} \tag{8.78}$$

Figure 8.3 shows the normalized mean density given by (8.78) for $\rho(t) = 10^3$ and $\eta = 0; 1; 2$.

8.4.6 *Finite Lifetime of the Economy and Transition to the Power Law Regime*

In reality, any economy has a finite lifetime, which leads to interesting questions on the transient regime before the establishment of the power law tail of the normalized mean density of firm's sizes $\mathcal{G}(\kappa, \alpha, \eta, \rho, \tau)$. Here, we study the dependence of

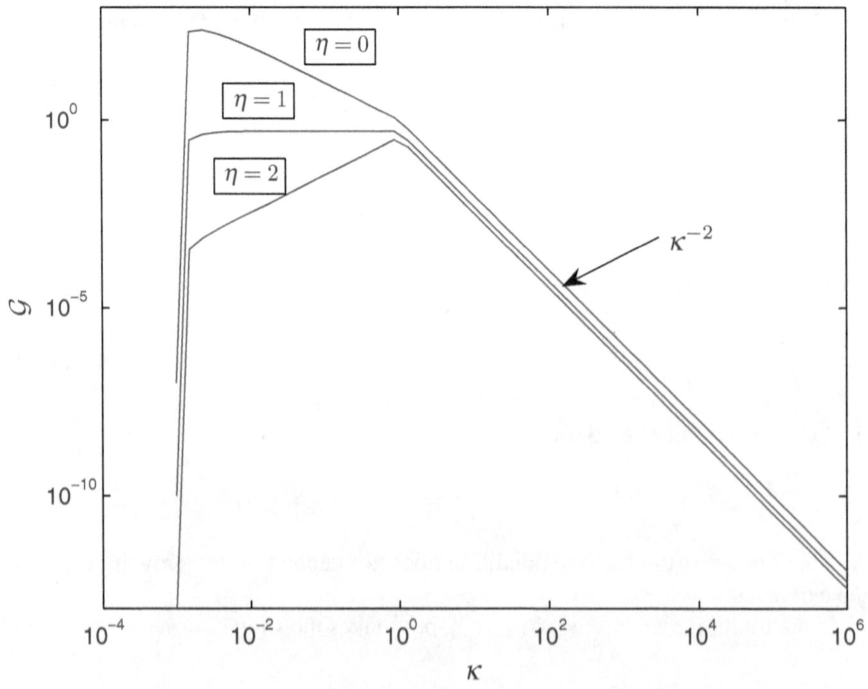

Fig. 8.3 Plot of the limit normalized density given by (8.78), when the balance condition holds, for $\rho = 10^3$ and $\eta = 0; 1; 2$

$\mathcal{G}(\kappa, \alpha, \eta, \rho, \tau)$ on the (dimensionless) age $\tau = \frac{b^2}{2}\theta_0$ of the economy. For this, let us first give the complete analytical expression of $\mathcal{G}(\kappa, \alpha, \eta, \rho, \tau)$:

$$\begin{aligned}\mathcal{G}(\kappa, \alpha, \eta, \rho, \tau) =& \frac{1}{2\alpha(\eta)}\kappa^{\frac{\alpha}{2}-1}\Bigg[e^{-\frac{1}{2}\alpha(\eta)|\ln\kappa|}\text{erfc}\left(\frac{|\ln\kappa| - \tau\alpha(\eta)}{2\sqrt{\tau}}\right) \\ &- e^{\frac{1}{2}\alpha(\eta)|\ln\kappa|}\text{erfc}\left(\frac{|\ln\kappa| + \tau\alpha(\eta)}{2\sqrt{\tau}}\right) \\ &- (\kappa\rho^2)^{-\alpha(\eta)/2}\text{erfc}\left(\frac{\ln(\kappa\rho^2) - \tau\alpha(\eta)}{2\sqrt{\tau}}\right) \\ &+ (\kappa\rho^2)^{\alpha(\eta)/2}\text{erfc}\left(\frac{\ln(\kappa\rho^2) + \tau\alpha(\eta)}{2\sqrt{\tau}}\right)\Bigg]. \quad (8.79)\end{aligned}$$

It is easy to check that this expression reduces to (8.76) in the limit $\tau \to +\infty$.

The complete formula (8.79) allows us to visualize how the normalized mean density of firm's sizes converges to the asymptotic power law (8.76). Figure 8.4 shows the normalized mean density (8.79) for the balanced case $\alpha = \eta - 1$, for $\eta = 3$, $\rho = 10^2$ and for two ages of the economy: $\tau = 1$ and $\tau = \infty$. It is

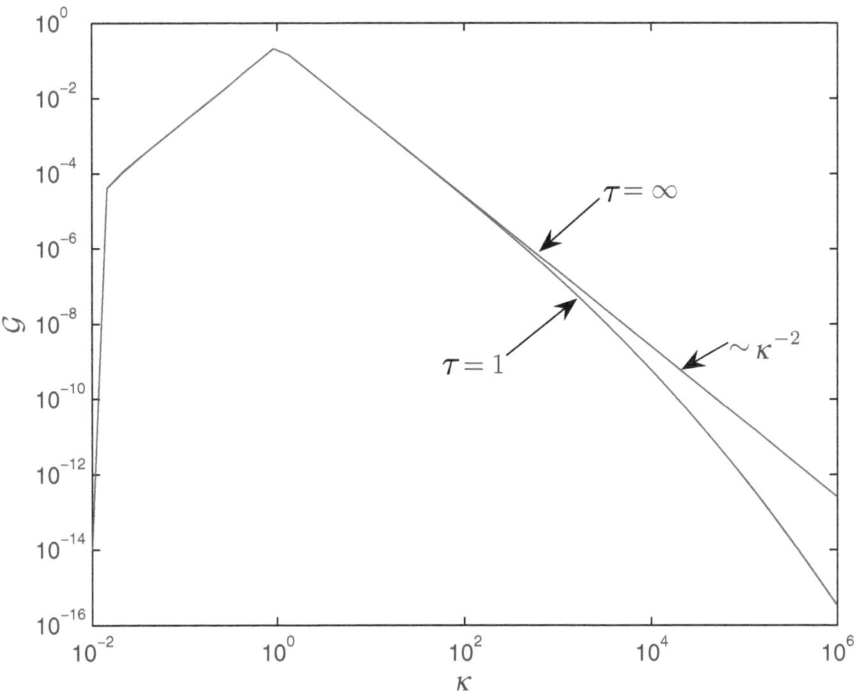

Fig. 8.4 Plot of the normalized mean density given by (8.78), satisfying the balance condition, for $\rho = 10^2$, $\eta = 3$, and two ages of the economy: $\tau = 1$ and $\tau = \infty$

interesting to see that, even for quite small ages of the economy, Zipf's law holds to a good approximation over three decades. At the same time, a clear deviation from a pure power law can be observed in the tail, typically for $s \gtrsim 10^4 \cdot s_0$. This deviation from Zipf's law reflects the limited lifetime of the economy in which firms have not had time yet to grow to very large sizes. The deviation of expression (8.79) from a pure power law provides the potential to extract some meaningful information on the parameters that would be otherwise hidden in the universal character of Zipf's law, such as the dimensionless relative drift $\alpha = \lambda - \lambda_0 = \frac{2}{b^2}(c - c_0)$, the composed birth and death hazard rates $\eta = \frac{2}{b^2}(d + \mu)$ and the standard deviation b of the firm growth stochastic process.

We now proceed to a quantitative determination of the deviations from the power law tail (8.76) due to the finite age τ of the economy. For this, it is convenient to study the s-dependence of the mean number of firms whose sizes exceed a given level s:

$$N(s,t) = \int_s^{\infty} g(s', t) ds'. \tag{8.80}$$

Zipf's law corresponds to $N(s,t) \sim s^{-1}$ for large s. As before, we study the normalized mean number of firms of normalized sizes larger than κ:

$$\mathcal{N}(\kappa, \alpha, \eta, \rho, \tau) := \int_{\kappa}^{\infty} \mathcal{G}(\kappa', \alpha, \eta, \rho, \tau) d\kappa'. \tag{8.81}$$

While it is straightforward to calculate this integral for any $\kappa > 0$, we restrict our analysis here to the range $\kappa > 1$ ($s > s_0(t)$).

Expression (8.79), for $\kappa > 1$, can be rewritten in the more convenient form:

$$\mathcal{G}(\kappa, \alpha, \eta, \rho, \tau) = \frac{1}{2\alpha(\eta)\,\kappa} \left(A_-\, \kappa^{-m_-} + A_+\, \kappa^{m_+} \right), \tag{8.82}$$

where

$$\begin{aligned} A_- &:= \text{erfc}\left(\frac{\ln \kappa - \tau\alpha(\eta)}{2\sqrt{\tau}} \right) - \rho^{-\alpha(\eta)} \text{erfc}\left(\frac{\ln(\kappa\rho^2) - \tau\alpha(\eta)}{2\sqrt{\tau}} \right), \\ A_+ &:= \rho^{\alpha(\eta)} \text{erfc}\left(\frac{\ln(\kappa\rho^2) + \tau\alpha(\eta)}{2\sqrt{\tau}} \right) - \text{erfc}\left(\frac{\ln \kappa + \tau\alpha(\eta)}{2\sqrt{\tau}} \right). \end{aligned} \tag{8.83}$$

Substituting (8.82), (8.83) into (8.81), and using the table integral

$$\begin{aligned} \int_y^{\infty} \text{erfc}(\beta \ln x + \gamma) \frac{dx}{x^{\alpha+1}} = & \frac{1}{\alpha} \Bigg[y^{-\alpha} \text{erfc}(\beta \ln y + \gamma) - \exp\left(\frac{\alpha(\alpha + 4\beta\gamma)}{4\beta^2} \right) \\ & \times \text{erfc}\left(\beta \ln y + \gamma + \frac{\alpha}{2\beta} \right) \Bigg], \end{aligned} \tag{8.84}$$

we obtain

$$\mathcal{N}(\kappa, \alpha, \eta, \rho, \tau) = B_-\, \kappa^{-m_-} + B_+\, \kappa^{m_+} - C\,, \tag{8.85}$$

where ϱ_- and ϱ_+ are defined in (8.77) and

$$B_- := \frac{1}{2\alpha(\eta) m_-} \left[\text{erfc}\left(\frac{\ln \kappa - \tau\alpha(\eta)}{2\sqrt{\tau}} \right) - \rho^{-\alpha(\eta)} \text{erfc}\left(\frac{\ln(\kappa\rho^2) - \tau\alpha(\eta)}{2\sqrt{\tau}} \right) \right],$$

$$B_+ := \frac{1}{2\alpha(\eta) m_+} \left[\text{erfc}\left(\frac{\ln \kappa + \tau\alpha(\eta)}{2\sqrt{\tau}} \right) - \rho^{\alpha(\eta)} \text{erfc}\left(\frac{\ln(\kappa\rho^2) + \tau\alpha(\eta)}{2\sqrt{\tau}} \right) \right],$$

$$C := \frac{1}{2\eta} e^{-\eta\tau} \left[\text{erfc}\left(\frac{\ln \kappa - \tau\alpha}{2\sqrt{\tau}} \right) - \rho^{-\alpha} \text{erfc}\left(\frac{\ln(\kappa\rho^2) - \tau\alpha}{2\sqrt{\tau}} \right) \right].$$

Figure 8.5 shows the mean cumulative number $\mathcal{N}(\kappa, \alpha, \eta, \rho, \tau)$ of firms as a function of the normalized firm size κ, for $\alpha = -1$ and $\eta = 0$ satisfying the balance condition, for $\rho = 100$ and $\tau = 5; 10; 50$. As expected, the older the economy,

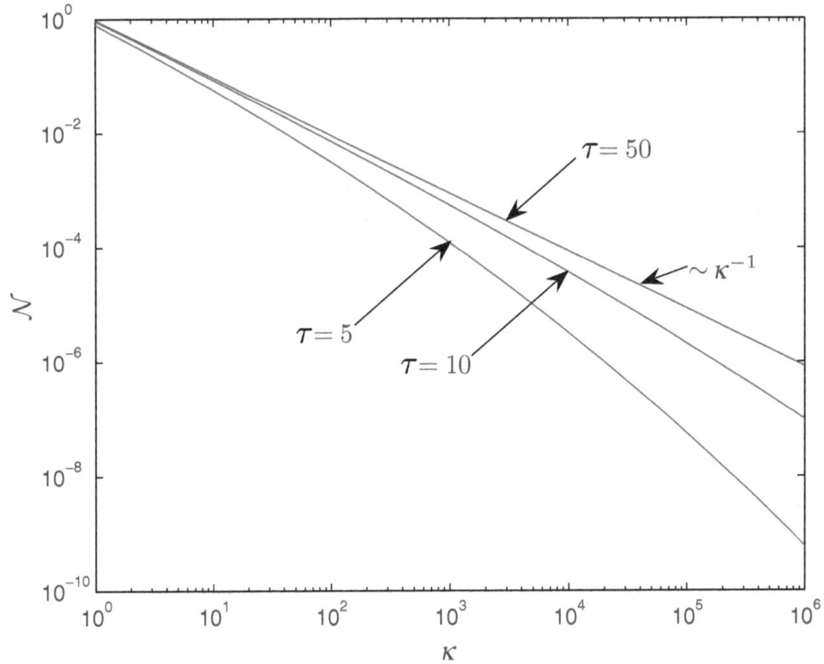

Fig. 8.5 Plots of mean number $\mathcal{N}(\kappa, \alpha, \eta, \rho, \tau)$ dependence on κ, for parameters $\alpha = -1$ and $\eta = 0$, satisfying to balance condition, for $\rho = 100$ and $\tau = 5$, 10, 50

the closer is the mean cumulative number $\mathcal{N}(\kappa, \alpha, \eta, \rho, \tau)$ to Zipf's law $\mathcal{N} \sim \kappa^{-1}$. Already for $\tau = 50$, $\mathcal{N}(\kappa, \alpha, \eta, \rho, \tau)$ obeys Zipf's law in the range $\kappa \in (1, 10^6)$.

Finally, by the same derivations as in the previous chapters, one can show that all the results discussed in this section hold in the case where the initial firm size factor s_0 is a random variable as long as the distribution of the initial firm sizes is not too fat-tailed, i.e., its moment of order m is finite.

8.5 Time-Dependence of the Average Size of the Global Economy of Firms

The economy considered here at a given time t is constituted of the firms that have been born in the past and are still in operation. It is interesting to ask what our model of firms, which includes birth, death and stochastic growth, predicts for the time-dependence of the total size of this economy. The derivation below leads to classify three regimes:

1. Stationarity of the real economy
2. Balanced growth
3. Positive growth

This calculation will also provide a natural economic interpretation of the second balance condition (8.75) discussed earlier for Zipf's law to hold.

In order to describe the time-dependence of the size of the global economy, we have to account for (1) the increase (or decrease) of the number of active firms due to the entry and exit processes and (2) the collective aggregate of the growth (or decay) of the sizes of the existing firms under their governing stochastic multiplicative growth processes. Including these two ingredients, the mean size of the global economy of firms can be written as

$$\Omega(t) := \int_0^t \nu(\zeta) \mathcal{Q}(t,\zeta) \mathrm{E}[S(t,\zeta)] d\zeta, \qquad (8.86)$$

where $\nu(t)$ is the birth intensity, $\mathcal{Q}(t,\zeta)$ is the survival function taking into account the sudden firm exits occurring at random times, $S(t,\zeta)$ is the size at time t of some firm born at a previous time ζ and $\mathrm{E}[S(t,\zeta)]$ its statistical average over the stochastic growth process. In expression (8.86), we have taken for convenience of notations that the economy started at $t = 0$ so that the current time t is also the age of the economy.

In what follows, we still suppose that the birth intensity is exponentially growing with time, while the survival function is exponentially decaying with time:

$$\nu(t) := \nu e^{dt}, \qquad \mathcal{Q}(t,\zeta) := e^{-\mu(t-\zeta)}. \qquad (8.87)$$

Then, expression (8.86) reads

$$\Omega(t) = \int_0^t \nu e^{-d\zeta}\, e^{-\mu(t-\zeta)} \mathrm{E}[S(t,\zeta)] d\zeta, \qquad (8.88)$$

In a first step, we assume that there is no minimum exit level (or equivalently, it is zero), and that the current firm size $S(t,\zeta)$ follows a pure geometric Brownian motion

$$S(t,\zeta) = s_0(\zeta)\, e^{c(t-\zeta)+bW(t-\zeta)}, \qquad (8.89)$$

where $W(\cdot)$ is a standard Wiener process. This implies that

$$\mathrm{E}[S(t,\zeta)] = s_0(\zeta)\, e^{a(t-\zeta)}, \qquad \text{where} \qquad a = c + \frac{b^2}{2}. \qquad (8.90)$$

Substituting this expectation into (8.88), and taking the initial size of firms at creation to grow exponentially with time (so as to reflect the presence of inflation for instance),

$$s_0(\zeta) := s_0\, e^{c_0 \zeta}, \qquad (8.91)$$

we obtain

$$\Omega(t) = \int_0^t \nu e^{d\zeta}\; e^{-\mu(t-\zeta)}\; s_0 e^{c_0\zeta + a(t-\zeta)} d\zeta. \qquad (8.92)$$

After simple transformations, expression (8.92) can be written in the form

$$\Omega(t) = s_0(t)\,\nu(t)\,\mathcal{T}(t,\gamma), \tag{8.93}$$

where

$$\mathcal{T}(t,\gamma) := \int_0^t e^{\gamma v}\,dv = \frac{1}{\gamma}\left(e^{\gamma t}-1\right), \tag{8.94}$$

and

$$\gamma := a - c_0 - d - \mu. \tag{8.95}$$

Expression (8.93) shows that the growth or decay of the nominal size of the economy is driven by three factors:

1. The size $s_0(t)$ of firms at birth, which grows in nominal terms to reflect the existence of inflation
2. The intensity $\nu(t)$ of new firms created in the economy, which reflects several factors, including pro-business legislation and tax laws, entrepreneur spirit as well as human population growth
3. The term $\mathcal{T}(t,\gamma)$ reflecting the overall success of typical firms in the economy

The sign of γ determines three qualitatively different growth regimes of the economy.

First Scenario ($\gamma < 0$), Stationary Real Economy per Firm

Expression (8.94) can be rewritten as

$$\mathcal{T}(t,\gamma) = \frac{1}{|\gamma|}\left(1-e^{-|\gamma|t}\right) \to \frac{1}{|\gamma|} \quad \text{for large } t. \tag{8.96}$$

Thus, at long times, the mean nominal size of the economy,

$$\Omega(t) = \frac{1}{|\gamma|} s_0(t)\,\nu(t) \sim e^{(c_0+d)t}, \tag{8.97}$$

is only driven by inflation and the intensity of firm entries. Since the number of active firms at time t is $\int_0^t \nu(t')dt' \sim e^{dt}$ at large times, we conclude that, in real terms (i.e., discounting for inflation) and per firm ($\sim$ per capita), the economy is stationary.

Second Scenario ($\gamma = 0$), Balanced Growth

Expression (8.94) leads to

$$\mathcal{T}(t,\gamma) = t, \tag{8.98}$$

so that the mean size of the economy is

$$\Omega(t) = s_0(t)\,\nu(t)\,t \sim t\,e^{(c_0+d)t}. \tag{8.99}$$

In real terms and per firm, the economy is growing linearly with time, corresponding to an asymptotically vanishing real growth rate per firm ($\sim$ per capita). This scenario for $\gamma = 0$ corresponds to the balanced condition $a - c_0 - d - \mu = 0$, that ensures the validity of Zipf's law, as demonstrated in Chap. 7.

Third Scenario ($\gamma > 0$), Growing Economy

Expression (8.94) shows that the real size of the economy grows with the real growth rate $\gamma + d + c_0$. The coefficient γ measures directly the contribution to the growth resulting from increasing performance of individual firms, for instance due to productivity gains.

Using the approach developed in Sect. 8.4, appendix shows that these results remain valid in the presence of a growing minimum size [given by expression (8.33)] below which firms exit.

Appendix

Influence of a Minimum Firm Size on the Classification of Economic Growth Regimes

Using the approach developed in Sect. 8.4, this appendix shows that the results obtained in Sect. 8.5 remain valid in the presence of a growing minimum size [given by expression (8.33)] below which firms exit.

In Sect. 8.4, we showed that, in the presence of a minimum firm size $s_1(t) = s_1 e^{c_1 t}$, the stochastic size of a given firm can be expressed in the form

$$S(t, u) = s_1(t)\, e^{\mathcal{Z}(t,\theta)}, \tag{8.100}$$

where

$$\mathcal{Z}(t, \theta) = z_*(\zeta) + (c - c_1)\theta + bW(\theta)\,, \qquad \theta = t - \zeta. \tag{8.101}$$

We have also determined the pdf $\varphi(z; t, \theta)$ of the auxiliary Wiener process with drift $\mathcal{Z}(t, \theta)$ under the form (8.47) that we recall for convenience,

$$\begin{aligned}\varphi(z; t, \theta) = &\frac{1}{2\sqrt{\pi\tau}}\left[\exp\left(-\frac{(z - \ln\rho(t) - \alpha\tau)^2}{4\tau}\right)\right.\\ &\left. - \rho^{-\alpha}(t)\exp\left(-\frac{(z + \ln\rho(t) - \alpha\tau)^2}{4\tau}\right)\right],\end{aligned} \tag{8.102}$$

where

$$\alpha = \frac{2}{b^2}\,(c - c_0)\,, \qquad \tau = \frac{b^2}{2}(t - \zeta)\,, \qquad \rho = \frac{s_0(t)}{s_1(t)}. \tag{8.103}$$

This allows us to calculate the expectation $\mathrm{E}[S(t,\zeta)]$ that we need for the determination of the mean size $\Omega(t)$ of the economy as given by (8.86):

$$\mathrm{E}\left[S(t,\zeta)\right] = s_1(t) \int_0^{\infty} e^z \, \varphi(z;t,t-\zeta) dz. \tag{8.104}$$

The calculation of this integral yields

$$\mathrm{E}\left[S(t,\zeta)\right] = s_0(t)\, e^{(1+\alpha)\tau} \Phi(\tau), \tag{8.105}$$

where

$$\begin{aligned} \Phi(\tau,\alpha,\rho) =& 1 - \frac{1}{2}\left[\operatorname{erfc}\left(\frac{(2+\alpha)\tau + \ln\rho}{2\sqrt{\tau}}\right)\right. \\ & \left. + \rho^{-2-\alpha}\operatorname{erfc}\left(\frac{-(2+\alpha)\tau + \ln\rho}{2\sqrt{\tau}}\right)\right]. \end{aligned} \tag{8.106}$$

The function $\Phi(\tau,\alpha,\rho)$ has the following limit

$$\lim_{\tau\to\infty} \Phi(\tau,\alpha,\rho) = \begin{cases} 1-\rho^{-2-\alpha}, & \alpha > -2, \\ 0 & \alpha < -2. \end{cases} \tag{8.107}$$

Reporting this limit in (8.105), we obtain the asymptotic dependence of the mean firm size

$$\mathrm{E}\left[S(t,\zeta)\right] \simeq s_0(t) e^{(1+\alpha)\tau}(1-\rho^{-2-\alpha}), \qquad \tau\to\infty, \qquad \alpha > -2. \tag{8.108}$$

This expression differs only by a positive constant factor from that obtained in the absence of the exit level ($\rho = \infty$). Thus, the classification of the three growth regimes obtained in Sect. 8.5 remains valid in the presence of a growing minimum size below which firms exit given by expression (8.33), as long as $\alpha > -2$. The condition $\alpha > -2$ holds for most the cases of interest, since the balance condition $\gamma = a - c_0 - d - \mu = 0$ translates into $\alpha = \frac{2(d+\mu)}{b^2} - 1$. Since the hazard rate h is positive, and as long as the intensity $\nu(t)$ is not decreasing ($d \geqslant 0$), we have $\alpha \geqslant -1$.

Chapter 9
Properties of the Realization Dependent Distribution of Firm Sizes

This chapter discuss the properties of the realization-dependent density of firm's sizes. We put together the different ingredients introduced in the previous chapters to analyze the extent to which the *mean* density of firm's sizes, which has been the main topic of this book, is representative of the *realized* density of firm's sizes in a given economy (i.e., a single realization).

9.1 Derivation of the Poissonian Distribution of the Number of Firms

As a preparation to the analysis of the typical fluctuations decorating the time-dependence of the mean size of the global economy, we derive in this first section the distribution of the number of firms in the economy and its evolution as a function of time, taking into account the birth and the two death processes discussed in the previous chapters.

We use the following assumptions:

- The birth flow of firms, described by the set of birth dates $\{t_\ell\}$, is a Poissonian process with intensity $\nu(t)$, which is a continuous function of t.
- The firms sizes $\{S_\ell(t, t_\ell)\}$, for any given sequence of birth dates $\{t_\ell\}$, are mutually statistically independent stochastic processes.
- The sequence $\{t_\ell\}$ of birth dates does not depend on the random firm sizes $\{S_\ell(t, t_\ell)\}$.

It is convenient to represent the time interval $(-\infty, t]$ as a sequence of adjacent intervals

$$\mathcal{T}_k := (t - (k+1)\Delta, t - k\Delta] , \qquad k = 0, 1, 2, \ldots \tag{9.1}$$

For simplicity and without loss of generality (as Δ can be taken arbitrary and, in particular, to recover the continuous limit $\Delta \to 0$), we assume that, in each k-th interval, the mean rate of firm births is the constant denoted as $\nu(t - k\Delta)$. This staircase representation of the intensity can be thought of as a discretized approximation to the assumed continuous function $\nu(t)$. To obtain the distribution of the number of firms, it is convenient to calculate the characteristic function

A. Saichev et al., *Theory of Zipf's Law and Beyond*,
Lecture Notes in Economics and Mathematical Systems 632,
DOI 10.1007/978-3-642-02946-2_9,

$$\Theta_k(u,s,t) := \mathrm{E}\left[e^{iu\tilde{N}_k(s,t)}\right] \tag{9.2}$$

of the random number

$$\tilde{N}_k(s,t) := \sum_{\ell:t_\ell\in\mathcal{T}_k} \mathbf{1}(S(t,t_\ell)-s), \tag{9.3}$$

of firms whose size is larger than s at time t and that were born in the interval $\mathcal{T}_k$.

In order to calculate the characteristic function $\Theta_k(u,s,t)$ (9.2), let us denote by M_k the random number of firms, that were born in the interval $\mathcal{T}_k$. Due to the Poissonian nature of the birth process, the probability that M_k is equal to m is given by

$$\Pr\{M_k = m\} = \frac{(\nu(t-k\Delta)\Delta)^m}{m!}\, e^{-\nu(t-k\Delta)\Delta}. \tag{9.4}$$

Correspondingly, the characteristic function $\Theta_k(u,s,t)$ defined by (9.2) is equal to

$$\Theta_k(u,s,t) = e^{-\nu(t-k\Delta)\Delta}\sum_{m=0}^{\infty}\frac{(\nu(t-k\Delta)\Delta)^m}{m!} \times \mathrm{E}\left[\exp\left(iu\sum_{\ell:t_\ell\in\mathcal{T}_k}\mathbf{1}(S(t,t_\ell)-s)\right)\Big|M_k=m\right]. \tag{9.5}$$

The expectation $\mathrm{E}[\cdot]$ is taken over the statistics of firms sizes $S(t,t_\ell)$ and over the m random birth dates $\{t_\ell\}$ which are statistically independent and uniformly distributed within the interval $\mathcal{T}_k$. Taking into account that all $\{S(t,t_\ell)\}$'s are statistically independent, and assuming that the statistical properties of any given firm size $S(t,t_\ell)$ at time t are stationary (i.e., remain the same for any $t_\ell\in\mathcal{T}_k$), we obtain

$$\mathrm{E}\left[\exp\left(iu\sum_{\ell:t_\ell\in\mathcal{T}_k}\mathbf{1}(S(t,t_\ell)-s)\right)\Big|M_k=m\right] = \mathrm{E}\left[e^{iu\mathbf{1}(S(t,t-k\Delta)-s)}\right]^m. \tag{9.6}$$

This allows us to transform (9.5) into

$$\Theta_k(u,s,t) = \exp\left(\nu(t-k\Delta)\Delta\left[\mathrm{E}\left[e^{iu\mathbf{1}(S(t,t-k\Delta)-s)}\right]-1\right]\right), \tag{9.7}$$

where the expectation $\mathrm{E}[\cdot]$ corresponds to averaging over the statistics of the random size $S(t,t-k\Delta)$ of a firm.

In order to calculate the expectation in (9.7), we use the identity valid for any real value y:

$$e^{iu\mathbf{1}(y)} = 1 + (e^{iu}-1)\mathbf{1}(y), \tag{9.8}$$

in which $\mathbf{1}(y)$ is the indicator function. Accordingly

$$\mathrm{E}\left[e^{iu\cdot\mathbf{1}(S(t,t-k\Delta)-s)}\right] - 1 = (e^{iu}-1)\mathrm{E}\left[\mathbf{1}(S(t,t-k\Delta)-s)\right]. \tag{9.9}$$

We thus obtain

$$\Theta_k(u,s,t) = \exp\left[(e^{iu}-1)\nu(t-k\Delta)\Delta\,\bar{F}(s;t,t-k\Delta)\right], \tag{9.10}$$

where we have used the definition,

$$\bar{F}(s;t,t-k\Delta) := \mathrm{E}\left[\mathbf{1}(S(t,t-k\Delta)-s)\right], \tag{9.11}$$

of the complementary cumulative distribution function (ccdf) $\bar{F}(s;t,t_0)$ of the current size $S(t,t_0)$ of some firm, which was born at the instant t_0. In other words, $\bar{F}(s;t|s_0,t_0) = \int_s^\infty f(s';t|s_0,t_0)ds'$, where $f(s;t|s_0,t_0)$ is the pdf of the size $S(t,t_0)$ of some firm.

Given the statistical independence of the firm sizes $\{S(t,t_\ell)\}$ and the statistical independence of the Poissonian birth dates of firms within non-overlapping intervals, the characteristic function of the total number $\tilde{N}(s,t)$ of firms whose sizes are larger than s is equal to the product of the characteristic functions of the numbers of firms which were born within the intervals $\mathcal{T}_k$:

$$\Theta(u,s,t;\Delta) = \prod_{k=0}^{\infty}\Theta_k(u,s,t) = \exp\left[(e^{iu}-1)N(s,t;\Delta)\right], \tag{9.12}$$

where

$$N(s,t;\Delta) := \sum_{k=0}^{\infty}\Delta\nu(t-k\Delta)\bar{F}(s;t-k\Delta). \tag{9.13}$$

By taking the continuous limit $\Delta \to 0$, we finally obtain the expression of the characteristic function $\Theta(u,s,t) = \lim_{\Delta\to 0}\Theta(u,s,t;\Delta)$ of the random number $\tilde{N}(s,t)$ of firms with sizes larger s:

$$\Theta(u,s,t) = \exp\left[(e^{iu}-1)N(s,t)\right], \tag{9.14}$$

where

$$N(s,t) := \mathrm{E}\left[\tilde{N}(s,t)\right] = \lim_{\Delta\to 0}N(s,t;\Delta) = \int_0^t \nu(t-\xi)\bar{F}(s;t,t-\xi)d\xi \tag{9.15}$$

is the mean number of firms of sizes larger than s at time t. The expression (9.14) for the characteristic function of the random number $\tilde{N}(s,t)$ implies

Proposition 9.1.1. *Under the assumptions made at the begin of this section, the random number $\tilde{N}(s,t)$ of firms whose size is larger than s in a given economy follows a Poisson law with parameter $N(s,t)$, i.e., the probability that $\tilde{N}(s,t)$ is equal to n is*

$$\Pr\{\tilde{N}(s,t) = n\} = \frac{[N(s,t)]^n}{n!}\, e^{-N(s,t)}. \tag{9.16}$$

Remark 9.1.1. Expression (9.15) for the mean current number $N(s,t)$ of firms of sizes larger than s remains valid even if the processes governing the firm's sizes $\{S(t,t_\ell)\}$ are mutually statistically dependent and the birth flow $\{t_\ell\}$ is not Poissonian. A necessary and sufficient condition for formula (9.15) to hold is that the stochastic process $S(t,t_\ell)$, describing the size of a firm born at instant t_ℓ, be statistically dependent solely on its own birthdate t_ℓ and not on any other element of the sequence $\{t_\ell\}$. However, the validity of formula (9.15) does not imply necessarily that the random number $\tilde{N}(s,t)$ is Poissonian. For this later property to hold, one needs in addition the size processes $\{S(t,t_\ell)\}$ to be mutually statistically independent and the birthdates $\{t_i\}$ to be Poissonian as well, as mentioned above.

Remark 9.1.2. The Poissonian property of the random number $\tilde{N}(s,t)$ is robust with respect to different ingredients impacting the lives of the firms. This is the case with respect to the two exit mechanisms discussed in Chap. 5 (by dropping below a minimum level) and in Chap. 7 (by a sudden exit at some random time), for which the random number $\tilde{N}(s,t)$ remains Poissonian.

Each exit mechanism requires its specific treatment. For the first one treated in Chap. 5, we need to solve the diffusion equation for the probability density function (pdf) of firm sizes, defined by

$$f(s;t,t_\ell) = -\frac{\partial \bar{F}(s;t,t_\ell)}{\partial s}, \tag{9.17}$$

in the presence of an absorbing boundary condition at the exit level.

For the sudden exit mechanism of Chap. 7, provided that its occurrence does not depend on the firm size, one can account for it explicitly in the formalism leading to (9.14) with (9.15) by rewriting relation (9.15) in the form

$$N(s,t) = \int_0^t H(t,t-\xi)\nu(t-\xi)\bar{F}(s;t,t-\xi)d\xi, \tag{9.18}$$

where $H(t,t_\ell)$ is the survival function of the firm born at t_ℓ. If the sudden exit times are statistically stationary, then $H(t,t_\ell) = H(t-t_\ell)$. Assuming in addition a Poissonian statistics for the sudden exit times implies that the survival function is exponential,

$$H(t,t_\ell) = e^{-\mu(t-t_\ell)}, \qquad t > t_\ell, \tag{9.19}$$

where μ is the hazard rate. In this case, expression (9.18) reduces to

$$N(s,t) = \int_0^t \nu(t-\xi)e^{-\mu\xi}\bar{F}(s;t,t-\xi)d\xi. \tag{9.20}$$

9.2 Finite-Size and Statistical Fluctuation Effects on the Empirical Measurement of Zipf's Law

Expression (9.16) shows that the distribution of firm sizes $\tilde{N}(s,t)$ is Poissonian. It allows us to quantify the typical deviations from the pure mathematical Zipf's law, that occur especially in the tail for large firm sizes s for which the number of firms in any empirical sample becomes limited.

As a metric of the statistical fluctuations occurring in the tail of empirical distributions of firm sizes due to finite-size effects, we consider the ratio of the realization-dependence number of firms of size larger than s at time t, $\tilde{N}(s,t)$, to its statistical average $N(s,t)$. Its variance reads

$$\text{Var}\left[\frac{\tilde{N}(s,t)}{N(s,t)}\right] = \frac{\text{E}\left[\tilde{N}^2(s,t) - N^2(s,t)\right]}{N^2(s,t)}. \tag{9.21}$$

We use the well-known result that all cumulants κ_m of a Poissonian random variable are all identical and equal to

$$\kappa_m = N(s,t)\ , \qquad m = 1,\ 2, \ldots \tag{9.22}$$

In particular

$$\text{Var}[\tilde{N}(s,t)] = N(s,t) \tag{9.23}$$

so that

$$\text{Var}\left[\frac{\tilde{N}(s,t)}{N(s,t)}\right] = \frac{1}{N(s,t)}. \tag{9.24}$$

To give a quantitative illustration, let us consider firms whose sizes evolve according to the pure geometric Brownian motion with no minimum exit size. Then,

$$N(s,t) = N(s) = \int_s^\infty g(s')ds', \tag{9.25}$$

where

$$g(s) = \frac{2\nu}{b^2|\lambda|}\, s_0^{-\lambda}\, s^{\lambda-1}\ , \qquad s > s_0\ , \qquad \lambda < 0. \tag{9.26}$$

This leads to

$$N(s) = N_0 \left(\frac{s_0}{s}\right)^{-\lambda}, \tag{9.27}$$

where

$$N_0 = \int_{s_0}^\infty g(s)ds = \frac{2\nu}{b^2\lambda^2} \tag{9.28}$$

is the mean number of firms, whose sizes, at a given time t, are larger than the initial size s_0. Using (9.24) and (9.27), we obtain

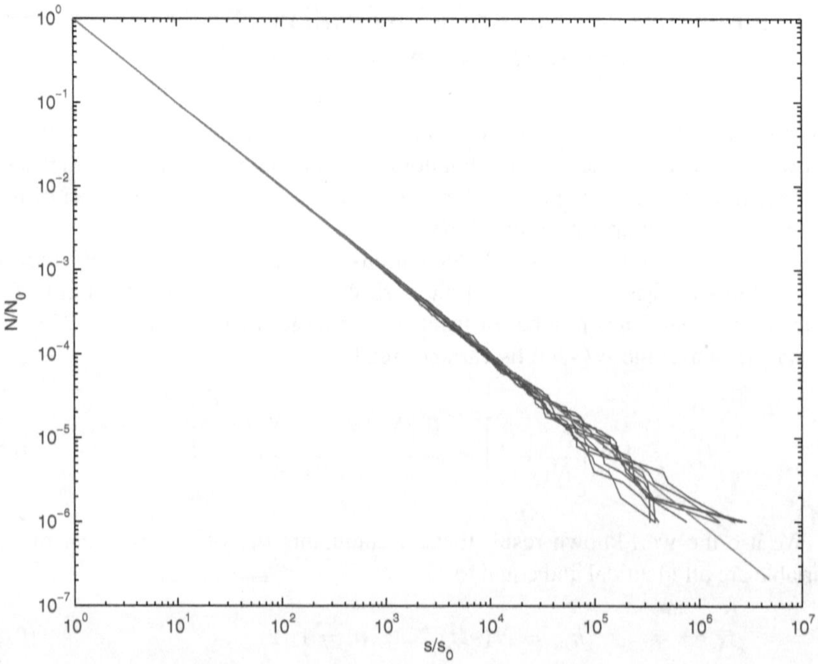

Fig. 9.1 Number of firms whose size is larger than s when $\sigma = 0.01$, $\nu = 50$, $\lambda = -1$ for ten realizations of the economy. The *straight line* depicts Zipf's law for the *mean* number of firms

$$\mathrm{Var}\left[\frac{\tilde{N}(s,t)}{N(s,t)}\right] = \frac{1}{N_0}\,\frac{s}{s_0}, \tag{9.29}$$

where we have assumed that $\lambda = -1$, so that Zipf's law $N(s) \sim s^{-1}$ should hold exactly in absence of statistical fluctuations.

In this illustrative example, the total number of firms is infinite while N_0 remains finite. Let us consider a data set spanning the range $s \in [s_0, s_*]$ where $s_* := 0.01\, N_0\, s_0$ is such that $\mathrm{Var}\left[\frac{\tilde{N}(s,t)}{N(s,t)}\right]$ remains smaller than 10^{-2} over this entire range. Suppose that the mean number of firms in the economy, whose sizes are larger than s_0, is equal to $N_0 = 10^6$. Then $s_* = 10^4\, s_0$, showing that Zipf's law should be observed with good accuracy over four orders of magnitudes in this example. Figure 9.1 depicts ten simulation results obtained for such an economy.

9.3 Estimation of the Distribution of Firm Sizes

Let us now discuss some specificities of the estimation of the distribution of firm sizes. For simplicity, we focus on the simple case where there is no exit level and where both the firm's birth intensity ν and the initial size of entrant firms s_0 are constant. We first consider the mean density of firm's size $g(s)$ for $s \geq s_0$. Recall that in such a case the mean number of firms whose sizes are larger than s is

$$N(x) = N_0 \, x^{\delta - 1} \,, \qquad x := \frac{s}{s_0}, \tag{9.30}$$

where N_0 is the mean number of firms, whose size is larger than s_0. Let us divide the semi-axis $x \in [1, \infty)$ into adjacent intervals

$$x \in [x_k, x_{k+1}) \,, \qquad x_k = \Delta^{k-1} \,, \qquad \Delta > 1 \,, \qquad k = 1, 2, \ldots \tag{9.31}$$

The mean number of firms, whose sizes belong to the kth interval, is equal to

$$N_k := N(x_k) - N(x_{k+1}) = A(\Delta, \lambda) N(x_k) \,, \qquad A(\Delta, \lambda) := 1 - \Delta^{\lambda}. \tag{9.32}$$

Then, plotting the points of coordinates $\{(y_k, z_k)\}$, with

$$z_k := \log_{10} N_k \,, \qquad y_k := \log_{10} x_k, \tag{9.33}$$

we obtain an approximate graph of the mean density of firm's size in a double logarithmic scale, for which Zipf's Law is characterized by a straight line with slope equal to -1.

Given a random sample of firms sizes $\tilde{S}_1, \tilde{S}_2, \ldots$, we can plot on the same graph the points of coordinates $\{(y_k, \tilde{z}_k)\}$, with

$$\tilde{z}_k := \log_{10} \tilde{N}_k, \tag{9.34}$$

where $\{\tilde{N}_k\}$ are random numbers of firms, whose sizes lie within the interval $[x_k, x_{k+1})$. It is easy to show that, under the assumptions in Sect. 9.1, all the random numbers $\{N_k\}$ are independent and distributed according to a Poisson law with mean and variance

$$\mathrm{E}\left[\tilde{N}_k\right] = \mathrm{Var}\left[\tilde{N}_k\right] = N_k. \tag{9.35}$$

If, additionally, $N_k \gtrsim 10^2$, one may interpret the random numbers $\tilde{N}_k$ as Gaussian, with very high accuracy. We can then estimate the number k_* of intervals, for which the mean numbers of firms are larger than some given N_*. It follows from (9.30) and (9.32) that

$$k_* = \frac{1}{|\lambda| \log_{10} \Delta} \left[\log_{10} \left(\frac{N_0}{N_*} \right) + \log_{10} \left(\Delta^{-\lambda} - 1 \right) \right]. \tag{9.36}$$

Taking here, for instance

$$N_0 = 10^6 \,, \qquad N_* = 10^2 \,, \qquad \Delta = 0 \,, \qquad \Delta = 10^{1/8}, \tag{9.37}$$

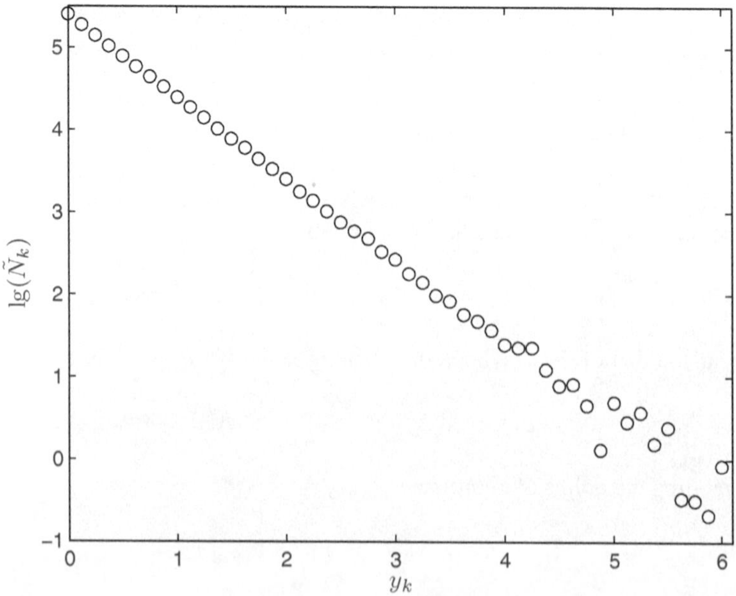

Fig. 9.2 Empirical estimates of Zipf's law for values of $\tilde{N}_k$ obtained by numerical simulation with parameter values $N_0 = 10^6$, $N_* = 10^2$, $\lambda = -1$ and $\Delta = 10^{1/8}$

we obtain

$$k_* \approx 36.$$

This means that, over more than four decades, i.e., from $y_1 = 0$ to $y_{k_*} \approx \frac{36}{8} = 4.375$, the relative error between the estimated density and the mean density of firms sizes, given by $\frac{\mathrm{E}[\tilde{N}_k]}{\sqrt{\mathrm{Var}[\tilde{N}_k]}} \leq \frac{1}{\sqrt{N_*}}$, is at most 10%. Figure 9.2 illustrates this fact by depicting the set of numerically simulated points (9.34), for the parameter values (9.37). As expected, we observe that Zipf's law is estimated with a very high accuracy over the interval $x \in [1, 10^4]$.

9.4 Statistical Fluctuations of the Size of the Global Economy Using Characteristic Functions

This section extends the analysis of Sect. 8.5 by characterizing the typical fluctuations decorating the time-dependence of the mean size of the global economy. We use the method of characteristic functions and the reasonings developed in the previous Sect. 9.1.

Let us consider a specific realization of the economy, whose nominal size at time t is the sum of the sizes of all existing firms:

$$\tilde{\Omega}(t) := \sum_{\ell} S(t, t_\ell). \tag{9.38}$$

Following the same reasoning that led to formula (9.14) for the characteristic function of the current number of firms, the characteristic function $\Psi(u,t) = \mathrm{E}\left[e^{iu\Omega(t)}\right]$ of the total current size of the economy is given by

$$\Psi(u,t) = e^{\mathcal{L}(u,t)}, \tag{9.39}$$

where

$$\mathcal{L}(u,t) = \int_0^t \nu(\zeta) H(t,\zeta) \left(\mathrm{E}\left[e^{iuS(t,\zeta)}\right] - 1\right) d\zeta. \tag{9.40}$$

These expressions allow us to obtain the cumulants of the distribution of the size of the economy at a given time t over many equivalent statistical realizations. Indeed, by definition, the cumulants $\kappa_n(t)$ are the coefficients of the power series expansions with respect to iu of the logarithm of the characteristic function:

$$\mathcal{L}(u,t) = \sum_{n=1}^{\infty} \kappa_n(t) \frac{(iu)^n}{n!}. \tag{9.41}$$

Expanding $\mathrm{E}\left[e^{iuS(t,\zeta)}\right]$ in powers of iu, expression (9.40) yields

$$\kappa_n(t) = \int_0^t \nu(\zeta) H(t,\zeta) \mathrm{E}[S^n(t,\zeta)] d\zeta. \tag{9.42}$$

The first four cumulants give respectively the mean size $\kappa_1(t)$ of the economy, its variance $\kappa_2(t)$, its skewness $\kappa_3(t)/[\kappa_2(t)]^{3/2}$ and its excess kurtosis $\kappa_4(t)/[\kappa_2(t)]^2$. We thus recover expression (8.86) for the mean size of the economy. The variance of the fluctuations of the size of the economy at time t is

$$\mathrm{Var}\left[\tilde{\Omega}(t)\right] = \int_0^t \nu(\zeta) H(t,\zeta) \mathrm{E}[S^2(t,\zeta)] d\zeta. \tag{9.43}$$

For simplicity, let us consider the case where the exit level is zero (no exit due to small firm sizes), while the intensity of firm's birth, the hazard function and the initial size at birth are exponential functions of time described by the relations (8.87) and (8.91). We also assume that the firm size processes $S(t,\zeta)$ are pure geometric Brownian motions given by (8.89). This leads to

$$\mathrm{E}[S^n(t,\zeta)] = s_0^n(t) \exp\left[n\left(c - c_0 + \frac{n}{2} b^2\right)(t - \zeta)\right], \tag{9.44}$$

so that the cumulants $\kappa_n(t)$ are given by

$$\kappa_n(t) = \nu(t)\, s_0^n(t)\, \mathcal{T}(t, \gamma_n), \tag{9.45}$$

where $\mathcal{T}(t,\gamma)$ is defined by (8.94), and

$$\gamma_n = n\left[\gamma + \frac{1}{2}(n-1)b^2\right], \quad \text{with} \quad \gamma = a - c_0 - d - \mu\,. \tag{9.46}$$

The main consequence of this calculation is that, according to the classification of the economic regime made in Sect. 8.5, even for stationary economies ($\gamma < 0$) for which the mean $\mathcal{T}(t,\gamma)$ tends to a finite limit as $t \to \infty$, there always exists a finite value

$$n_* = \frac{2|\gamma|}{b^2} < \infty \tag{9.47}$$

such that, for $n > n_*$, $T(t,\gamma_n)$ grows exponentially without bounds at long times. Thus, even in a stationary economy, cumulants of order $n > n_*$ grow without limit, expressing the existence of ever increasing statistical fluctuations decorating the mean size of the economy. Of course, this phenomenon holds true with even stronger amplitude for $\gamma \geqslant 0$.

For illustration, let us consider the balance case $\gamma = 0$ for which

$$\gamma_n = \frac{1}{2}n(n-1)b^2. \tag{9.48}$$

In this case,

$$\begin{aligned}
\kappa_1(t) &\simeq \tau\,\tilde{\nu}(t)\,s_0(t)\,, \quad \text{with} \quad \tilde{\nu}(t) := \frac{2}{b^2}\,\nu(t)\,, \quad \tau := \frac{b^2}{2}\,t, \\
\kappa_n(t) &\simeq \frac{\tilde{\nu}(t)}{n(n-1)}\,s_0^n(t)\,e^{n(n-1)\tau}\,, \quad n > 1\,.
\end{aligned} \tag{9.49}$$

These expressions allow us to explore the validity of (1) the law of large numbers (LLN) and (2) the central limit theorem (CLT). The LLN describes how the sample mean of a random variable converges to its expected value. This convergence (for the so-called "strong law") is controlled by the dimensionless ratio $\kappa_2(t)/[\kappa_1(t)]^2$. In the balanced case $\gamma = 0$, the asymptotic behavior of this parameter is

$$\frac{\kappa_2(t)}{[\kappa_1(t)]^2} \simeq \frac{e^{2\tau}}{2\tau^2\tilde{\nu}(t)} \sim e^{(b^2-d)t}. \tag{9.50}$$

Thus, for $d < b^2$, i.e., if the proportional fluctuations describing the stochastic growth of individual firms are sufficiently large compared with the growth rate the firm entry rate, then the size of the global economy becomes more and more random with the course of time and the LLN is not satisfied.

A necessary condition for the validity of the CLT is that the excess kurtosis $\kappa_4(t)/[\kappa_2(t)]^2$ tends asymptotically to zero. Using the expressions in (9.49) for the balanced case $\gamma = 0$, we obtain the following asymptotic dependence for the excess kurtosis of the size of the global economy:

$$\frac{\kappa_4(t)}{[\kappa_2(t)]^2} \simeq \frac{e^{8\tau}}{3\tilde{\nu}(t)} \sim e^{(4b^2-d)t}. \tag{9.51}$$

This indicates that, for $d < 4b^2$, the distribution of the size of the global economy over many equivalent statistical realizations becomes less and less Gaussian at a function of time and the CLT does not hold. For $b^2 < d < 4b^2$, the LLN holds but the CLT does not. Gabaix (2005) has built on this intuition to suggest that idiosyncratic firm-level fluctuations can explain an important part of aggregate shocks, and provide a microfoundation for aggregate productivity shocks.

Chapter 10
Future Directions and Conclusions

10.1 Mergers and Acquisitions and Spin-offs

The set of mechanisms that we have considered up to now explains quite well both ends of the distribution of firm's sizes. Indeed, while the average growth rate of firm's asset values, the intensity of firm's births and the hazard rate of a firm's sudden death accurately explain the behavior of the population of large firms, the introduction of a lower threshold below which firms disappear (by lack of efficiency, for instance) explains the behavior of the population of small firms. Everything is not perfect however. In many countries, like France and India amongst others, the population of medium size firms exhibits an anomalous behavior. One usually speaks of the "missing middle" phenomenon to describe the fact there is a deficit of firms of intermediate sizes. Our model, while already quite versatile, does not account for this country-specific stylized fact. The reason is in fact rather simple. To a large extent, the "missing middle" phenomenon could be explained by the strong propensity of large firms to merge with firms which are still small but present a promising potential. This should yield a depletion of the population of medium size firms.

In addition to the mechanisms in terms of birth, death and random growth which have been considered in the previous chapters, we envision that the next level of development of a complete mathematical theory of the dynamics of the population of firms needs to take into account the mechanism of mergers between firms (referred to as M&A for "merger and acquisition"), as well as it symmetric, the phenomenon of creation of spin-off firms born from parent firms which outsource a part of their existing business as separate units. For this, the long tradition in physics concerning the investigation of the processes of coagulation (merger) and of fragmentation (spin-off) could provide a fertile reservoir of ideas and techniques (Aldous, 1999; Leyvraz, 2003).

10.1.1 General Formalism

The approach that we have used up to now, in which each firm follows its own path of stochastic growth and death, does not apply anymore. Indeed, due to both

A. Saichev et al., *Theory of Zipf's Law and Beyond*,
Lecture Notes in Economics and Mathematical Systems 632,
DOI 10.1007/978-3-642-02946-2_10, © Springer-Verlag Berlin Heidelberg 2010

the M&A and the spin-off processes, firms are not independent from one another. Therefore, the distribution of firm's sizes $g(s,t)$ is not simply given by the integral equation (3.15) but by an integro-differential equation that expresses the coupling between firms introduced by M&A and spinoffs. This equation can be decomposed into three main contributions:

- As we have seen, in the absence of M&A and of spin-offs, $g(s,t)$ is solution to

$$\left[\frac{\partial g(s,t)}{\partial t}\right]_{\text{int. growth}} = -\frac{\partial}{\partial s}\left(a(s,t)g(s,t)\right) + \frac{1}{2}\frac{\partial^2}{\partial s^2}\left(b^2(s,t)g(s,t)\right) + \delta(s-s_0)\cdot \nu(t), \tag{10.1}$$

with $g(s;0) = 0$ if the intensity of firm's birth $\nu(t)$ equals zero for all $t < 0$, provided that the density $f(s;t)$ of the single firm's size satisfies

$$\frac{\partial f(s;t)}{\partial t} + \frac{\partial}{\partial s}\left(a(s,t)f(s;t)\right) = \frac{1}{2}\frac{\partial^2}{\partial s^2}\left(b^2(s,t)f(s;t)\right), \tag{10.2}$$

with $f(s;0) = \delta(s-s_0)$. These equations account for the creation of firms and their subsequent stochastic growth as explained in Sect. 2.7 and following. In expression (10.1), we have omitted the hazard rate of firm's sudden death as well as the lower threshold s_1 below which firms disappear as they can be easily reintroduced.
- Firm's growth via mergers and acquisitions can be accounted for by the term

$$\left[\frac{\partial g(s,t)}{\partial t}\right]_{\text{M\&A}} = \frac{1}{2}\int_0^s M(s-s',s')g(s-s';t)g(s',t)ds' - g(s,t)\int_0^\infty M(s,s')g(s';t)ds', \tag{10.3}$$

where $M(s,s') \geqslant 0$ is the rate of mergers between pairs of firms of sizes s and s' to form a new firm of size $s+s'$. The first term in the r.h.s. of (10.3) describes the creation of a firm of size s by merging two firms of sizes s' and $s-s'$. The second term in the r.h.s. of (10.3) states that the population of firms of size s decreases when some of them merge with any other firm, thereby creating a larger firm. We neglect here any dilution of size that may occur during the merger.
- The opposite of M&A is the process of *spin-off* in which a firm parts with a fraction of its assets to create a new firm. This process is accounted for by

$$\left[\frac{\partial g(s,t)}{\partial t}\right]_{\text{spin-off}} = -g(s,t)\int_0^s SO(s-s',s')ds' + 2\int_0^\infty SO(s,s')g(s+s',t)ds', \tag{10.4}$$

where $SO(s, s') \geq 0$ denotes the rate of spin-off by which a firm of size $s + s'$ creates a new firm endowed with a part s' of its *pre*-spin-off asset and retains an asset value equals to s. The first term in the r.h.s. of (10.4) describes the process in which a firm of size s decreases from size s to size $s - s'$ while creating a new spin-off firm of size $s' < s$. The second term in the r.h.s. of (10.4) accounts for the creation of firms of size s as spin-offs of large firms.

Putting all the terms together, the dynamic of the density of firm sizes is solution to the integro-differential equation

$$\begin{aligned} \frac{\partial g(s,t)}{\partial t} &= \frac{1}{2} \int_0^s M(s-s',s') g(s-s',t) g(s',t) ds' \\ &\quad - g(s,t) \int_0^\infty M(s,s') g(s',t) ds' - g(s,t) \int_0^s SO(s-s',s') ds' \\ &\quad + 2 \int_0^\infty SO(s,s') g(s+s',t) ds' - \frac{\partial}{\partial s} \left(a(s,t) g(s,t) \right) \\ &\quad + \frac{1}{2} \frac{\partial^2}{\partial s^2} \left(b^2(s,t) g(s,t) \right) + \delta(s-s_0) \cdot \nu(t). \end{aligned} \tag{10.5}$$

Specific instances of this class of integro-differential equations has been studied to describe the processes of coagulation (merger) and of fragmentation (spin-off) in the physics and mathematical literature (Aldous, 1999; Leyvraz, 2003). Here, given the fact that the stochastic growth process is included, there are not general solutions. The specific solutions exist under simplifying assumptions on the functional dependence of the rates of M&A (M) and of spin-off creation (SO), while often neglecting the stochastic growth process (i.e., putting $a(s,t) = b(s,t) = 0$).

In order to make further progress, we first make the additional assumption that both M and SO are constants, independent of the firm sizes. Consequently, the previous equation simplifies into

$$\begin{aligned} \frac{\partial g(s,t)}{\partial t} &= \frac{M}{2} \int_0^s g(s-s',t) g(s',t) ds' - M \cdot g(s,t) \int_0^\infty g(s',t) ds' \\ &\quad - SO \cdot s \cdot g(s,t) + 2SO \int_0^\infty g(s+s',t) ds' - \frac{\partial}{\partial s} \left(a(s,t) g(s,t) \right) \\ &\quad + \frac{1}{2} \frac{\partial^2}{\partial s^2} \left(b^2(s,t) g(s,t) \right) + \delta(s-s_0) \cdot \nu(t). \end{aligned} \tag{10.6}$$

10.1.2 *Mergers and Acquisitions and Spin-offs with Brownian Internal Growth*

We first consider the case where the internal growth dynamics (10.1) is an arithmetic Brownian motion, i.e., a and b are constants. Besides, we assume that

$$g'(0) \equiv \lim_{s\to 0+} \frac{dg(s)}{ds} < \infty. \tag{10.7}$$

Under the assumption that ν does not depend on t explicitly, we look for the stationary solution of (10.6), and focus on the large s behavior of the solution(s) $g(s)$ that satisfies the boundary conditions $g(0) = 0$ and $g(s) \to 0$ as s goes to infinity. This condition (10.7) can be justified by the "absorption" conditions for small firms which disappear below a small size.

The above conditions allow us to perform the Laplace transform of (10.6), which yields

$$\frac{M}{2}\hat{g}(k)^2 - \left[M\hat{g}(0) + ak - \frac{b^2}{2}k^2\right]\hat{g}(k) - \frac{b^2}{2}g'(0) \\ +SO\cdot\left[\partial_k\hat{g}(k) - 2\frac{\hat{g}(k)-\hat{g}(0)}{k}\right] + \nu e^{-ks_0} = 0, \tag{10.8}$$

where

$$\hat{g}(k) := \int_0^\infty g(s)e^{-ks}ds. \tag{10.9}$$

This equation is of the Ricatti type. We can therefore perform the change of function

$$\hat{g}(k) = \frac{2\cdot SO}{M}\frac{F'(k)}{F(k)}, \tag{10.10}$$

so that (10.8) reads

$$F''(k) + \left[\frac{b^2k^2}{2\cdot SO} - \frac{ak}{SO} - \frac{M}{SO}\hat{g}(0) - \frac{2}{k}\right]F'(k) \\ + \left[\frac{M\hat{g}(0)}{SO\,k} + \frac{M\nu e^{-ks_0}}{2SO^2} - \frac{Mb^2g'(0)}{4SO^2}\right]F(k) = 0. \tag{10.11}$$

It is a second order ordinary differential equation with a regular singular point at $k = 0$. It can be locally solved by the Frobenius method and, after a little algebra, we find that

$$\hat{g}(k) = \frac{M}{2SO}\hat{g}(0) + \left[\frac{M\nu}{2SO^2} - \frac{Mb^2g'(0)}{4SO^2} - \frac{M^2\hat{g}(0)^2}{4SO^2}\right]k \\ = \left[\frac{3M^2\hat{g}(0)\,\nu}{8SO^3} - \frac{3M^2\hat{g}(0)\,b^2g'(0)}{16SO^3} + \frac{M\nu\,s_0}{6SO^2}\right. \\ \left. - \frac{7M\hat{g}(0)\,a}{12SO^2} - \frac{M^3g(0)^3}{8SO^3} - \frac{b^2}{2SO} + C\right]k^2 \\ + \left[\frac{M\nu\,s_0}{2SO^2} + \frac{M\hat{g}(0)\,a}{2SO^2}\right]k^2\ln(k) + O(k^3), \tag{10.12}$$

where C is an arbitrary constant. A solution exists if the following consistency condition holds,

$$\frac{M}{2SO} = 1 \,, \tag{10.13}$$

which ensures that the r.h.s. of (10.12) recovers $\hat{g}(0)$ in the limit $k \to 0$. This consistency condition expressed a balance between firm's mergers and fragmentations through spin-offs. If this condition holds, we get

$$\begin{aligned}
\hat{g}(k) &= \hat{g}(0) + \left[\frac{\nu}{SO} - \frac{b^2 g'(0)}{2SO} - \hat{g}(0)^2\right] k \\
&= \left[\frac{3\hat{g}(0)\,\nu}{2SO} - \frac{3\hat{g}(0)\,b^2 g'(0)}{4SO} + \frac{\nu\, s_0}{3SO} - \frac{7\hat{g}(0)\,a}{6SO} - g(0)^3 - \frac{b^2}{2SO} + C\right] k^2 \\
&\quad + \left[\frac{\nu\, s_0}{SO} + \frac{\hat{g}(0)\,a}{SO}\right] k^2 \ln(k) + O(k^3),
\end{aligned} \tag{10.14}$$

where $\hat{g}(0)$ remains undetermined. By virtue of Theorem 8.1.6 in Bingham et al. (1987), $g(s)$ is regularly varying at infinity with a tail index m equal to 2, i.e., $g(s) \sim L(s)/s^{1+m}$ with $m = 2$ for large s, where $L(s)$ is a slowly varying function. This tail exponent is one unit more that the value for Zipf's law to hold. This section provides a novel potentially interesting mechanism for the emergence of a power law tail with exponent 2. Actually, Frobenius' method shows that the origin of the value 2 of the exponent can be traced back to the term proportional to $(2/k)F'(k)$ in (10.11), which itself comes from the term $(-2(\hat{g}(k) - \hat{g}(0))/k$ in (10.8), which comes from the spin-off term $2SO \int_0^\infty g(s + s')ds'$ in (10.6). This shows that the internal growth dynamics of firms does not a play a role in determining the exponent: intuitively, the arithmetic Brownian motion creates fluctuations in the sizes of firms which are small compared with those resulting from the spin-off mechanism.

Again, a stationary power law is obtained here when a balance condition (10.13) holds. The study of the different possible regimes appearing when (10.13) is violated is an interesting avenue for future research. Possible new regimes can appear, such as "shattering" into a universe dominated by firms of minimal size with one individual (dynamics dominated by spin-off) or the emergence of a monopolistic firm (dynamics dominated by mergers). These regimes have been studied in specific cases in previous investigations of the physics of coagulation-fragmentation processes (Leyvraz, 2003).

10.1.3 Mergers and Acquisitions and Spin-offs with GBM for the Internal Growth Process

Under the assumption that ν does not depend on t explicitly and that the internal growth follows a GBM ($a(s,t) = a \cdot s; b(s,t) = b \cdot s$), (10.6) for $g(s)$ becomes

$$\frac{M}{2}\int_0^s g(s-s')g(s')ds' - M\cdot g(s)\int_0^\infty g(s')ds' - SO\cdot s\cdot g(s)$$
$$+2SO\int_0^\infty g(s+s')ds' - a\cdot\frac{\partial\, s\cdot g(s)}{\partial s} + \frac{b^2}{2}\frac{\partial^2\, s^2\cdot g(s)}{\partial s^2} + \nu\cdot\delta(s-s_0) = 0. \tag{10.15}$$

The solution(s) $g(s)$ have to satisfy the boundary conditions $g(0) = 0$ and $g(s) \to 0$ as s goes to infinity.

Since the GBM process of firm's value growth, together with birth and death processes, can create Zipf's law with an exponent $m = 1$ smaller than the value obtained with the arithmetic Brownian motion in the presence of M&A and spin-offs, one can expect that the solution of this equation exhibits interesting competitions between the different mechanisms. Here, we just show the nature of the nonlinear nonlocal equation as a future challenge towards the development of a more complete theory of the dynamics of interacting firms.

10.2 Summary of Main Results

10.2.1 Importance of Balance Conditions for Zipf's Law

Starting from Gibrat's rule of proportionate growth, and considering some departures thereof, we have investigated the extent to which Zipf's law could be the robust result of the dynamics of a population of firms in an evolving economy. We have shown that taking into account the birth and death of firms, in addition to their random growth, does not qualitatively alter the shape of the distribution of firm's sizes. The typical power law shape of the distribution of firm's size is a pervasive feature. However, the tail index that characterizes the hyperbolic decay of the distribution does depend on several characteristics of the economic environment. Amongst others, the average growth rate of firm's asset values, the firm's birth intensity and the hazard rate of a firm's sudden death have a direct impact on the value of the tail index.

A general result unraveled by our study is that Zipf's law is strictly valid if and only if a balanced condition is fulfilled: the sum of all the mechanisms responsible for the growth and decline of firms must vanish on average in a precise sense. Any departure from this requirement yields a departure of the tail index from its canonical value $m = 1$. This result can allow one to understand why different tail indexes are reported in the literature for different countries around the world. However, the reasons that underpin the validity of the balance condition are not yet clear. No economic law can justify why all these mechanisms should almost exactly compensate one another.

In the absence of such economic arguments, one would be tempted to resort to Gabaix's explanation based upon the idea that, in order to make stationary the distribution of firm's sizes, one has to first remove the impact of the overall economy on the growth of each individual firm. Therefore, since the overall economy grows at the same rate as each individual firm, on average, the balance condition is satisfied in the referential of the growing economy.

However, this argument can be justified, and Gabaix (1999)'s derivation of Zipf's law holds, only for models of the economy in which all firms are born at the same time. In Gabaix (1999)'s framework, because the firms are all born at the same instant, firms grow – on average – at the same rate as the overall economy. Consequently, when discounted by the global growth rate of the economy, the average expected growth rate of the firms must be zero.

10.2.2 Essential Differences with Gabaix (1999)'s Derivation of Zipf's Law

Applied to the framework developed in this book, and focusing on the distribution of *discounted* firm sizes, Gabaix (1999)'s argument would lead to the conditions $a = \mu$, with $d = c_0 = c_1 = 0$ (in order to match Gabaix's assumptions). Gabaix (1999)'s condition would thus seem to be equivalent to our balance condition for Zipf's law describing the density of firms' sizes to hold. Actually, this reasoning is incorrect. Consider the case where $a > \mu$, such that the global economy grows at the average growth rate $r_G = a - \mu$. Gabaix (1999) proposed to measure the growth of a firm in the frame of the global economy. In this frame, the new average growth rate of the firm is $a' = a - r_G = \mu$, which indeed would suggest that the balance condition is *automatically* obeyed when a is replaced by a'. But, one should notice that a' is a transformed growth rate, and not the true rate. The average growth rate $r_G = a - \mu$ of the global economy is micro-founded on the contributions of all growing firms. It would be incorrect to insert a' in the statements of Proposition 7.4.1, as a' is the effective growth rate resulting from the change of frame, while our exact derivation requires the parameters a and μ for Proposition 7.4.1 to hold. As such, nothing in our model automatically sets the growth rate a of firms to their death rate μ, contrarily to what happens in Gabaix (1999)'s model. The main difference that invalidates the application of Gabaix (1999)'s argument is the stochastic flow of firm's births and deaths.

Summing up, Gabaix (1999)'s derivation of Zipf's law relies crucially on a model view of the economy in which *all firms are born at the same instant*. Our approach is thus essentially different since it considers the flow of firm births, as well as their deaths, which is more in agreement with empirical evidence.

In addition, it is important to stress that the available empirical evidence on Zipf's law is based on analyzing *cross-sectional* distributions of firm sizes, i.e., at specific times. As a consequence, the change to the global economic growth frame, argued by Gabaix (1999), just amounts to multiplying the value of each firm by the same constant of normalization, equal to the size of the economy at the time when the cross-section is measured. Obviously, this normalization does not change the exponent of the power law distribution of sizes, if it exists.

Furthermore, elaborating on Krugman (1996)'s argument about the non-convergence of the distribution of firm sizes toward Zipf's law in Simon (1955)'s model, Blank and Solomon (2000) have shown that Gabaix (1999)'s argument suffers from a more technical problem. Based on the demonstration that the two limits, the number of firms $N \to \infty$ and $s_{\min}/\mathrm{E}[S] \to 0$[1] (or equivalently the limit of large times $t \to \infty$) are non-commutative, Blank and Solomon (2000) showed that Zipf's exponent $m = 1$ as obtained by Gabaix (1999)'s argument requires (1) taking the long time limit $s_{\min}/\mathrm{E}[S] \to 0$ over which the economy made of a large but finite number N of firms grows without bounds, while simultaneously obeying the condition (2) $N \gg \exp[\mathrm{E}[S]/s_{\min}]$. The problem is that conditions (1) and (2) are mutually exclusive. Blank and Solomon (2000) showed that this inconsistency can be resolved by allowing the number of firms to grow proportionally to the total wealth of the economy.

As we have shown in Chaps. 7 and 8, our approach allows one to overcome Blank and Solomon (2000)'s objection to Gabaix's model. Indeed, in Gabaix's model, the balance condition ensures that the size of any firm goes to zero almost surely. Thus, since all the firms are born at the same time, it is necessary to introduce a repelling lower bound in order to let some of them keep a non-vanishing size. Otherwise, the size distribution degenerates. In contrast, our approach assumes an unremitting flow of firm's births. So, even if our balance condition implies that firm sizes almost surely go to zero (or to the lower admissible firm size) as in Gabaix's model, our economy always exhibits a significant number of firms whose size is arbitrarily large so that the convergence toward the limit distribution always occurs, whether or not Blank and Solomon's condition holds. This issue has been discussed in more details in Chap. 9, which studies the effect of the finite age of the economy on the shape of the distribution of firm sizes. In addition, since expression (8.69) with (8.68) for the normalized mean density of firm's sizes also takes into account the case where the rate of firms' births increases exponentially as $\nu(t) = \nu_0 e^{d \cdot t}$, when $d > 0$, our analysis provides a generalization of Blank and Solomon (2000)'s results.

10.2.3 *Robustness of Zipf's Law as an Attractor for Large Variance of the GMB of Firm's Growth*

Our theory suggests a simple explanation for the empirical evidence that the exponent m for the distribution of firm's sizes is close to 1 (Zipf's law): the exponent m of the tail of the distribution of firms sizes is a weakly varying function of the parameters ν, a, b, μ, as shown explicitly in Fig. 7.5, where one can observe that the exponent m is close to 1 for a very wide range of parameters. As a quantitative illustration, consider a globally growing economy $a > \mu$, with growth rate $r_G = a - \mu =$ 2–3% per year, and $\mu = 0.15$ per year, corresponding to about 50% of the firms disappearing within five years of their existence. Putting $c_0 = c_1 = 0$, this yields $\varepsilon = \mu/a = 0.88$. For this value, Fig. 7.5 shows that $0.9 \le m < 1$ for all values of $\delta' = b^2/2a$, which is empirically undistinguishable from the exact Zipf's law value $m = 1$ in available empirical data.

[1] The term $\mathrm{E}[S]$ refers to the statistical average over the finite but large population of N firms.

References

Abramovitz E, Stegun IA (1964) Handbook of Mathematical Functions (New York: Dover publications)

Adamic LA, Huberman BA (2000) The nature of markets in the World Wide Web. Q J Electron Commer 1:5–12

Adamic LA, Huberman BA (2002) Zipf's law and the Internet. Glottometrics 3:143–150

Aldous DJ (1999) Deterministic and stochastic models for coalescence (aggregation and coagulation): a review of the mean-field theory for probabilists. Bernoulli 5(1):3–48

Almus M, Nerlinger E (2000) Testing Gibrats law for young firms – empirical results for West Germany. Small Bus Econ 15:1–12

Aoyama H, Fujiwara Y, Souma W (2004) In: Takayasu H (eds) Proceedings of Second Nikkei Symposium on Econophysics. Springer, Tokyo

Audretsch D (1995) Innovation, growth and survival. Int J Ind Organ 13:441–457

Audretsch D, Santarelli E, Vivarelli M (1999) Start-up size and industrial dynamics: some evidence from Italian manufacturing. Int J Ind Organ 17:965–983

Axtell RL (2001) Zipf distribution of US firm sizes. Science 293:1818–1820

Barabasi A-L, Albert R (1999) Emergence of scaling in random networks. Science 286:509–512

Barabasi A-L, Albert R (2002) Statistical mechanics of complex networks. Rev Mod Phys 74: 47–97

Bartelsman E, Scarpetta S, Schivardi F (2003) Comparative analysis of firm demographics and survival: micro-level evidence for the OECD countries. OECD Economics Department Working Papers, No. 348. OECD. doi:10.1787/010021066480

Bateman G, Erdelyi A (1954) Tables of integral transforms, vol I. McGraw-Hill, New York

Becchetti L, Trovato G (2002) The determinants of growth for small and medium sized firms. The role of the availability of external finance. Small Bus Econ 19(4):291–306

Bingham NH, Goldie CM, Teugels JL (1987) Regular variation. Cambridge University Press, Cambridge

Blank A, Solomon S (2000) Power laws in cities population, financial markets and internet sites (scaling in systems with a variable number of com-ponents). Physica A 287(1–2):279–288

Bonaccorsi Di Patti E, Dell'Ariccia G (2004) Bank competition and firm creation. J Money Credit Banking 36:225–251

Borodin AN, Salminen P (2002) Handbook of Brownian motion – facts and formulae. Birkhäuser, Basel

Bottazzi G, Dosi G, Lippi M, Pammolli F, Riccaboni M (2001) Innovation and corporate growth in the evolution of the drug industry. Int J Ind Organ 19:1161–1187

Breaslau L, Cao P, Pan L, Phillips G, Shenker S (1999) Web caching and Zipf-like distributions: evidence and implications. IEEE INFOCOM'99, vol 1, pp 126–134

Brixy U, Grotz R (2007) Regional patterns and determinants of birth and survival of new firms in Western Germany. Entrepreneurship Reg Dev 19(4):293–312

Buldyrev SV, Pammolli F, Riccaboni M, Yamasaki K, Fu DF, Matia K, Stanley HE (2007a) Generalized preferential attachment model for business firms growth rates: II. Mathematical treatment. Eur Phys J B 57:131–138

Buldyrev SV, Growiec J, Pammolli F, Riccaboni M, Stanley HE (2007b) The growth of business firms: facts and theory. J Eur Econ Assoc 5(2–3):574–584

Cabral LMB, Mata J (2003) On the evolution of the firm size distribution: facts and theory. Am Econ Rev 93(4):1075–1090

Calvo JL (2006) Testing Gibrat's law for small, young and innovating firms. Small Bus Econ 26:117–123

Clauset A, Erwin DH (2008) The evolution and distribution of species body size. Science 321(5887):399–401

Daley DJ, Vere-Jones D (2007) An introduction to the theory of point processes. Springer, Berlin

Daniel G, Sornette D, Wohrmann P (2008) Look-ahead benchmark bias in portfolio performance evaluation. J Portfolio Manage Fall 2009. http://arXiv.org/abs/0810.1922

de Wit G (2005) Firm size distributions: an overview of steady-state distributions resulting from firm dynamics models. Int J Ind Organ 23:423–450

Duchin R, Levy M (2008) Portfolio optimization and the distribution of firm size. Working paper. http://papers.ssrn.com/sol3/papers.cfm?abstract_id=1146308

Dunne T, Roberts MJ, Samuelson L (1988) Patterns of firm entry and exit in US manufacturing industries. RAND J Econ 19(4):495–515

Dunne T, Roberts MJ, Samuelson L (1989) The growth and failure of US manufacturing plants. Q J Econ 104(4):671–698

Eeckhout J (2004) Gibrat's law for (all) cities. Am Econ Rev 94(5):1429–1451

Eckhout J (2009) Gibrats law for (all) cities: Reply, American Economic Review 99:1676–1683.

Embrechts P, Kluppelberg C, Mikosch T (1997) Modelling extremal events. Springer, Berlin

Fama EF, French KR (1993) Common risk factors in the returns on stocks and bonds. J Finance Econ 33:3–56

Fama EF, French KR (1995) Size and book-to-market factors in earnings and returns. J Finance 50:131–155

Feller W (1966) An introduction to probability theory and its applications, vol II. Wiley, New York

Fu D, Pammolli F, Buldyrev SV, Riccaboni M, Matia K, Yamasaki K, Stanley HE (2005) The growth of business firms: theoretical framework and empirical evidence. Proc Natl Acad Sci USA 102:18801–18806

Fujiwara Y (2004) Zipf law in firms bankruptcy. Physica A 337(1–2):219–230

Fujiwara Y, Di Guilmi C, Aoyama H, Gallegati M, Souma W (2004a) Do Pareto–Zipf and Gibrat laws hold true? An analysis with European firms. Physica A 335:197–216

Fujiwara Y, Aoyamac H, Di Guilmib C, Soumaa W, Gallegati M (2004b) Gibrat and Pareto–Zipf revisited with European firms. Physica A 344(1–2):112–116

Furusawa C, Kaneko K (2003) Zipf's law in gene expression. Phys Rev Lett 90:088102

Gabaix X (1999) Zipf's law for cities: an explanation. Q J Econ 114:739–767

Gabaix X (2005) The granular origins of aggregate fluctuations. Working paper. MIT and NBER

Gaffeo E, Gallegati M, Palestrini A (2003) On the size distribution of firms. Additional evidence from the G7 countries. Physica A 324:117–123

Galbraith JK (1997) The great crash, 1929. Houghton Mifflin, Boston

Glassman S (1994) A caching relay for the World Wide Web. Comput Netw ISDN Syst 27(2): 165–173

Gur Y, Nowak A, Rakocy K, Solomon S (2008) Microscopic study reveals the singular origins of growth (preprint). e-print at arXiv:0803.2201

Harhoff D, Stahl K, Woywode M (1998) Legal form, growth and exit for West German firms – results for manufacturing, construction, trade and service industries. J Ind Econ 46:453–488

Ijri Y, Simon HA (1977) Skew distributions and the sizes of business firms. North-Holland, New York

Ioannides YM, Gabaix X (2003) The evoluation of city size distributions. In: Henderson JV, Thisse J-F (eds) Handbook of urban and regional economics. North-Holland, Amsterdam

Kaizoji T, Iyetomi H, Ikeda Y (2006) Re-examination of the size distribution of firms. Evol Inst Econ Rev 2(2):183–198

Karatzas I, Shreve S (1991) Brownian motion and stochastic calculus, 2nd edn. Springer, Berlin

Karlin S, Taylor HM (1975) A first course in stochastic processes, vol I. Academic, New York

Kindleberger CP (2000) Manias, panics, and crashes: a history of financial crises, 4th edn. Wiley, New York

Kitov I (2009) What is the best firm size to invest? http://papers.ssrn.com/sol3/papers.cfm?abstract_id=1351731

Knaup AE (2005) Survival and longevity in the business employment dynamics data. Mon Labor Rev 2005(May):50–56

Kong JS, Sarshar N, Roychowdhury VP (2008) Experience versus talent shapes the structure of the Web. Proc Natl Acad Sci USA 105(37):13724–13729

Krugman P (1996) The self-organizing economy. Blackwell, Oxford

Lee Y, Amaral LAN, Canning D, Meyer M, Stanley HE (1998) Universal features in the growth dynamics of complex organizations. Phys Rev Lett 81:3275–3278

Levy M (2009) Gibrats law for (all) cities: Comment, American Economic Review 99:1672–1675.

Levy M, Solomon S (1996) Power laws are logarithmic Boltzmann laws. Int J Mod Phys C 7:595–601

Leyvraz F (2003) Scaling theory and exactly solved models in the kinetics of irreversible aggregation. Phys Rep 383:95–212

Lotti F, Santarelli E (2004) Industry dynamics and the distribution of firm sizes: a non-parametric approach. South Econ J 70(3):443–466

Lotti F, Santarelli E, Vivarelli M (2003) Does Gibrat's law hold among young, small firms? J Evol Econ 13(3):213–235

Lotti F, Santarelli E, Vivarelli M (2007) Defending Gibrat's Law as a long-run regularity. IZA Discussion Papers 2744. Institute for the Study of Labor (IZA)

Lucas RE Jr (1978) On the size distribution of business firms. Bell J Econ 9(2):508–523

Luttmer EGJ (2007) Selection, growth, and the size distribution of firms. Q J Econ 122(3): 1103–1144

Maillart T, Sornette D, Spaeth S, Von Krogh G (2008) Empirical tests of Zipf's law mechanism in open source Linux distribution. Phys Rev Lett 101:218701

Malcai O, Biham O, Solomon S (1999) Power-law distributions and Lévy-stable intermittent fluctuations in stochastic systems of many autocatalytic elements. Phys Rev E 60(2):1299–1303

Malevergne Y, Sornette D (2006) Extreme financial risks (from dependence to risk management). Springer, Heidelberg

Malevergne Y, Sornette D (2007) A two-factor asset pricing model and the fat tail distribution of firm sizes. Working paper. http://papers.ssrn.com/sol3/papers.cfm?abstract_id=960002

Malevergne Y, Pisarenko VF, Sornette D (2005) Empirical distributions of log-returns: between the stretched exponential and the power law? Quantitative Finance 5(4):379–401

Malevergne Y, Pisarenko VF, Sornette D (2009) Gibrat's law for cities: uniformly most powerful unbiased test of the Pareto against the lognormal (http://papers.ssrn.com/sol3/papers.cfm?abstract_id=1479481)

Morse PM, Feshbach H (1953) Methods of theoretical physics. McGraw-Hill, New York

Nielsen J (1997) Do websites have increasing returns? http://www.useit.com/alertbox/9704b.html

Nowak A, Kus M, Urbaniak J, Zarycki T (2000) Simulating the coordination of individual economic decisions. Physica A 287:613–630

Nowak A, Vallacher RR, Kus M, Urbaniak J (2005) The dynamics of societal transition: modeling nonlinear change in the Polish nonlinear system. Int J Sociol 35(1):65–88

Okuyama K, Takayasu M, Takayasu H (1999) Zipf's law in income distribution of companies. Physica A 269:125–131

Pammolli F, Fu D, Buldyrev SV, Riccaboni M, Matia K, Yamasaki K, Stanley HE (2007) Generalized preferential attachment model for business firms growth rates: I. Empirical evidence. Eur Phys J B 57:127–130

Podobnik B, Horvatic D, Pammolli F, Wang F, Stanley HE, Grosse I (2008) Size-dependent standard deviation for growth rates: empirical results and theoretical modeling. Phys Rev E 77:056102

Podobnik B, Horvatic D, Petersen AM, Stanley HE (2009) Quantitative relations between risk, return and firm size. Europhys Lett 85:50003. doi:10.1209/0295-5075/85/50003

Reid GC (1992) Early life cycle behaviour of micro-firms in Scotland. Small Bus Econ 7:89–95

Reynolds PD, Storey DJ, Weshead P (1994) Cross-national comparisons of the variation in new firm formation rates. Reg Stud 28:443–456

Riccabonia M, Pammolli F, Buldyrev SV, Pontac L, Stanley HE (2008) The size variance relationship of business firm growth rates. Proc Natl Acad Sci USA 105(50):19595–19600

Rossi-Hansberg E, Wright MLJ (2007a) Urban structure and growth. Rev Econ Stud 74(2):597–624

Rossi-Hansberg E, Wright MLJ (2007b) Establishment size dynamics in the aggregate economy. Am Econ Rev 97(5):1639–1666

Rozenfeld H, Rybski D, Andrade JS, Batty M, Stanley HE, Makse HA (2008) Laws of population growth. Proc Natl Acad Sci USA 105(48):18702–18707

Rozenfeld H, Rybski D, Gabaix X, Makse H (2009) Zipf's Law for all cities (preprint). Abstract at http://www.thic-apfa7.com/share/data/48_rozenfeld.pdf

Sakai K, Watanabe T (2009) The firms as a bundle of barcodes. Working Paper Series No. 38. JSPS Grants-in-Aid for Creative Scientific Research (Understanding Inflation Dynamics of the Japanese Economy). http://www.ier.hit-u.ac.jp/~ifd/doc/IFD_WP38.pdf

Schwarzkopf Y, Farmer JD (2008) Time evolution of the mutual fund size distribution. arXiv:0807.3800 (preprint)

Shefrin H (2000) Beyond greed and fear: understanding behavioral finance and the psychology of investing. Harvard Business School Press, Boston

Shiller RJ (2000) Irrational exuberance. Princeton University Press, Princeton, NJ

Shleifer A (2000) Inefficient markets: an introduction to behavioral finance. Oxford University Press, Oxford

Simon HA (1955) On a class of skew distribution functions. Biometrika 52:425–440

Simon HA (1960) Some further notes on a class of skew distribution functions. Information Control 3:80–88

Simon H, Bonini C (1958) The size distribution of business firms. Am Econ Rev 48(4):607–617

Sinha S, Pan RK (2006) How a "hit" is born: the emergence of popularity from the dynamics of collective choice. In: Chakrabarti BK, Chakraborti A, Chatterjee A (eds) Econophysics and sociophysics: trends and perspectives. Wiley-VCH, Berlin, ISBN-10: 3-527-40670-0

Sornette D (2003) Why stock markets crash: critical events in complex financial systems. Princeton University Press, Princeton, NJ

Sornette D (2006) Critical phenomena in natural sciences, chaos, fractals, self-organization and disorder: concepts and tools, 2nd edn. Springer Series in Synergetics. Springer, Heidelberg

Stanley MHR, Buldyrev SV, Havlin S, Mantegna RN, Salinger A, Stanley HE (1995) Zipf plots and the size distribution of firms. Econ Lett 49:453–457

Stanley MHR, Amaral LAN, Buldyrev SV, Haylin S, Leschhorn H, Maass P, Salinger MA, Stanley HE (1996) Scaling behavior in the growth of companies. Nature 379:804–806

Steindl J (1965) Random processes and the growth of firms. A study of the Pareto Law. Griffin, London

Sutton J (1997) Gibrat's legacy. J Econ Lit 35:40–59

Sutton J (2002) The variance of firm growth rates: the "scaling" puzzle. Physica A 312:577–590

Takayasu M, Sameshima S, Ohnishi T, Ikeda Y, Takayasu H, Watanable K (2008) Massive economics data analysis by econophysics methods – the case of companies' network structure. Epoth-making simulation. In: Annual Report of the Earth Simulator Center, April 2007 – March 2008. Earth Simulator Center, Yokohama, Chap. 3, pp. 263–268

Weiss CR (1998) Size, growth, and survival in the upper austrian farm sector. Small Bus Econ 10:305–312

White EN (ed) (1996) Stock market crashes and speculative manias. The International Library of Macroeconomic and Financial History, 13. Elgar, Cheltenham

Yamasaki K, Matia K, Buldyrev SV, Fu D, Pammolli F, Riccaboni M, Stanley HE (2006) Preferential attachment and growth dynamics in complex systems. Phys Rev E 74:035103(R)

Yanagita T, Onozaki T (2008) Dynamics of a market with heterogeneous learning agents. J Econ Interact Coord 3:107–118

Zipf GK (1949) Human behavior and the principle of least effort. Addison-Wesley, Cambridge, MA

Index

FSC
www.fsc.org
MIX
Papier | Fördert gute Waldnutzung
FSC® C083411

Kolibri 360 GmbH
Ferdinand-Jühlke-Straße 7
99095 Erfurt, Deutschland
produktsicherheit@kolibri360.de